T0361806

MATHEMATICAL AND EXPERIMENTAL MODELING OF PHYSICAL AND BIOLOGICAL PROCESSES

TEXTBOOKS in MATHEMATICS

Series Editor: Denny Gulick

PUBLISHED TITLES

COMPLEX VARIABLES: A PHYSICAL APPROACH WITH APPLICATIONS AND MATLAB®
Steven G. Krantz

INTRODUCTION TO ABSTRACT ALGEBRA
Jonathan D. H. Smith

LINEAR ALBEBRA: A FIRST COURSE WITH APPLICATIONS
Larry E. Knop

MATHEMATICAL AND EXPERIMENTAL MODELING OF PHYSICAL AND BIOLOGICAL PROCESSES
H. T. Banks and H. T. Tran

FORTHCOMING TITLES

ENCOUNTERS WITH CHAOS AND FRACTALS
Denny Gulick

TEXTBOOKS in MATHEMATICS

MATHEMATICAL AND EXPERIMENTAL MODELING OF PHYSICAL AND BIOLOGICAL PROCESSES

H. T. Banks

H. T. Tran

CRC Press
Taylor & Francis Group
Boca Raton London New York

CRC Press is an imprint of the
Taylor & Francis Group an **informa** business

A CHAPMAN & HALL BOOK

Chapman & Hall/CRC
Taylor & Francis Group
6000 Broken Sound Parkway NW, Suite 300
Boca Raton, FL 33487-2742

© 2009 by Taylor & Francis Group, LLC
Chapman & Hall/CRC is an imprint of Taylor & Francis Group, an Informa business

No claim to original U.S. Government works
Printed in the United States of America on acid-free paper
10 9 8 7 6 5 4 3 2 1

International Standard Book Number-13: 978-1-4200-7337-9 (Hardcover)

Library of Congress Cataloging-in-Publication Data

Banks, H. Thomas.
  Mathematical and experimental modeling of physical and biological processes
/ H.T. Banks, H.T. Tran.
    p. cm. -- (Textbooks in mathematics)
  Includes bibliographical references and index.
  ISBN 978-1-4200-7337-9 (hardcover : alk. paper)
  1. Science--Mathematical models. 2. Engineering--Mathematical models. I.
Tran, H. T. II. Title. III. Series.

  Q172.B36 2009
  501'.5118--dc22   2008049371

Visit the Taylor & Francis Web site at
http://www.taylorandfrancis.com

and the CRC Press Web site at
http://www.crcpress.com

Preface

For the past several years, the authors have developed and taught a two-semester modeling course sequence based on fundamental physical and biological processes: heat flow, wave propagation, fluid and structural dynamics, structured population dynamics, and electromagnetism. Among the specific topics covered in the courses were thermal imaging and detection, dynamic properties (stiffness, damping) of structures such as beams and plates, acoustics and fluid transport, size-structured population dynamics, electromagnetic dispersion and optics.

One of the major difficulties (theoretically, computationally, and technologically) in mathematical model development is the process of comparing models to the field data. Typically, mathematical models contain parameters and coefficients that are not directly measurable in experiments. Hence, experiments must be carefully designed in order to provide sufficient data for model parameters and/or coefficients to be determined accurately. In this context, a major innovative component of the course has been the exposure of students to specific laboratory experiments, data collection and analysis. As usual in such modeling courses, the pedagogy involves beginning with first principles in a physical, chemical or biological process and deriving quantitative models (partial differential equations with initial conditions, boundary conditions, etc.) in the context of a specific application, which has come from a *"client discipline"* — academic, government laboratory, or industrial research group, such as thermal nondestructive damage detection in structures, active noise suppression in acoustic chambers, smart material (piezoceramic sensing and actuation) structures vibration suppression, or optimizing the introduction of mosquitofish into rice fields for the control of mosquitos. The students then use the models (with appropriate computational software — some from MAT-LAB, some from the routines developed by the instructors specifically for the course) to carry out simulations and analyze experimental data. The students are exposed to experimental design and data collection through laboratory demos in certain experiments and through actual hands-on experience in other experiments.

Our experience with this approach to teaching advanced mathematics with a strong laboratory experience has been, not surprisingly, overwhelmingly positive. It is one thing to hear lectures on natural modes and frequencies (eigenfunctions and eigenvalues) or even to compute them, but quite another to go to the laboratory, *excite* the structure, *see* the modes, and *take* data to verify your theoretical and computational models.

Indeed, in writing this book, which is based on these experimentally oriented modeling courses, the authors aim to provide the reader with a fundamental understanding of how mathematics is applied to problems in science and engineering. Our approach will be through several "case study" problems that arise in industrial and scientific research laboratory applications. For each case study problem the perception on why a model is needed and what goals are to be sought will be discussed. The modeling process begins with the examination of assumptions and their translation into mathematical models. An important component of the book is the designing of appropriate experiments that are used to validate the mathematical model's development. In this regard, both hardware and software tools, which are used to design the experiments, will be described in sufficient detail so that the experiments can be duplicated by the interested reader. Several projects, which were developed by the authors in their own teaching of the above-mentioned modeling courses, will also be included.

The book is aimed at advanced undergraduate and/or first year graduate students. The emphasis of the book is on the application as well as what mathematics can tell us about it. The book should serve both to give the student an appreciation of the use of mathematics and also to spark student interest for deeper study of some of the mathematical and/or applied topics involved.

The completion of this text involved considerable assistance from others. Foremost, we would like to express our gratitude to many students, postdoctoral fellows and colleagues (university and industrial/government laboratory based scientists) over the past decades, who generously contributed to numerous research efforts on which our modules/projects are based. Specifically, we wish to thank Sarah Grove, Nathan Gibson, Scott Beeler, Brian Lewis, Cammey Cole, John David, Adam Attarian, Amanda Criner, Jimena Davis, Stacey Ernstberger, Sava Dediu, Clay Thompson, Zackary Kenz, Shuhua Hu and Nate Wanner among our many young colleagues for their assistance in reading various drafts or portions of this book. (Of course, any remaining errors, poor explanations, etc., are solely the responsibility of the authors.) Finally, the authors wish to acknowledge the unwavering support of our families in our efforts in the development and completion of this manuscript as well as other aspects of our professional activities. For their support, patience and love, this book is dedicated to Susie, John, Jennifer, Thu, Huy and Hoang.

H. T. Banks
H. T. Tran

List of Tables

List of Figures

Contents

Chapter 1

Introduction: The Iterative Modeling Process

We begin this monograph with a brief discussion of certain philosophical notions that are important in the modeling of physical and biological systems. Modeling in our view is simply a means for providing a conceptual framework in which real systems may be investigated. The modeling process itself is (or should be) most often an iterative process: one can distinguish in it a number of rather separate steps that usually must be repeated. This iterative modeling process is schematically depicted in Figure 1.1. One begins with the real system under investigation and pursues the following sequence of steps:

(i) empirical observations, experiments, and data collection;

(ii) formalization of properties, relationships and mechanisms that result in a *biological or physical model* (e.g., stoichiometric relations detailing pathways, mechanisms, biochemical reactions, etc., in a metabolic pathway model; stress-strain, pressure-force relationships in mechanics and fluids);

(iii) abstraction or mathematization resulting in a *mathematical model* (e.g., algebraic and/or differential equations with constraints and initial and/or boundary conditions);

(iv) formalization of uncertainty/variability in model and data resulting in a *statistical model* (this usually involves basic assumptions about errors in modeling, observation process/measurement, etc.);

(v) model analysis that can consist of simulation studies, analytical and qualitative analysis including stability analysis, and use of mathematical techniques such as perturbation studies, parameter estimation (inverse problems) data fitting, statistical analysis;

(vi) interpretation and comparison (with the real system) of the conclusions, predictions and conjectures obtained from step (v);

(vii) changes in "understanding" of mechanisms, pathways, etc., in the real system.

As one completes step (vii), one is led naturally to reformulate the physical or biological model by returning to either step (i) (if new experiments are indicated) or step (ii). In either case one then proceeds through the steps again, seeking to improve the findings of the previous transit through the sequence.

Steps (i), (ii), (iii), (iv) belong to what one might term the *formulation* stage of the modeling process, while step (v) is the *solution* stage of the modeling process, and steps (vi) and (vii) constitute the *interpretation* stage. In practice, however, it is often (unfortunately) the case that investigators do not make a clear distinction in the steps outlined here. This can lead to confusion and, in some cases, incorrect conclusions and gross misunderstanding of the real system.

Let us turn next to the reasons frequently given for modeling. Perhaps the one most often offered is *simplification*: the use of models makes possible the investigation of very complex systems in a systematic manner. A second rationale is *ease in manipulation*: investigations involving separation of subunits and hypothesis testing may often be facilitated through use of simulations in place of experimentation. The suggestive features in modeling can also help in *formulation of hypotheses* and in the *design of critical experiments*. The modeling process also requires *preciseness* in investigation in that one must move from a general, verbal explanation of phenomena to a specific, quantitative one.

But a rationale perhaps more fundamental than any of these is that modeling leads to an *organization* of inquiry in that it tends to polarize one's thinking and aid in posing basic questions concerning what one does and does not know for certain about the real system. Whatever the reasons that have been advanced to justify modeling attempts, it is sufficient perhaps to note that the primary goal must be *enlightenment*, that is, *to gain a better understanding of the real system*, and the success or lack thereof of any modeling attempt must be appraised with this in mind.

One must recognize the various *levels* or *multi-scale* aspects of modeling in any attempt to compare or assess the validity of several models for a phenomenon. For example, consider the phenomena involved in the transmission of a nerve impulse along an axon: this process is likely to be described by the mathematician or biophysicist in terms of partial differential equations, wave phenomena, or transmission line analogies, whereas a neurophysiologist might speak in terms of local circuit analogies and changes in conductances. The cell physiologist might describe the phenomena in the context of transport properties of membranes and ion flow, while the molecular biochemist could insist that the real story lay in the theory of molecular binding.

A second example involves the physical motion (vibration) of a structure such as a plate or beam. Again the mathematician might describe this in terms of a partial differential equation whereas the mechanical engineer might use a modal analysis (in terms of natural frequencies of oscillation) based on internal stress-strain relationships.

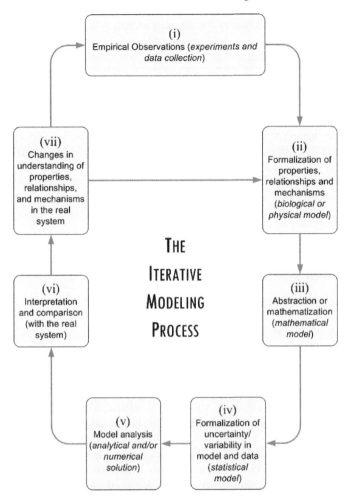

FIGURE 1.1: Schematic diagram of the iterative modeling process.

In each of the examples cited above, the different modeling approaches move to an increasingly more micro level. Each approach involves an attempt to explain a phenomenon that is not understood at one level by description at a more micro level (in general) where understanding is more complete. This attempt to explain "unknowns" in terms of more basic "knowns" is clearly the foundation of most modeling investigations. Indeed, in addition to noting that nerve impulse phenomena are described in terms of membrane conductances, permeabilities, ion flow, etc., one might observe that blood circulation is studied in the context of elementary hydrostatics and fluid dynamics while metabolic processes are usually investigated via use of the language of elementary chemical kinetics and thermodynamics.

The choice of the level (micro vs. macro) at which one models depends very much upon the training and background of the investigator. Furthermore, the perception of whether a model is a "good" one or not is also greatly influenced by this factor, and it is therefore not surprising that all of the approaches to the nerve impulse phenomena mentioned above (or indeed those for modeling any physical or biological phenomena) can be subjected to valid criticisms in any attempt to evaluate them.

Before discussing the criteria one might use in evaluating modeling investigations, let us list some of the common difficulties and limitations often encountered in the modeling of systems:

(a) Availability and accuracy of data;

(b) Analysis of the mathematical model;

(c) Use of local representations that are invalid for the overall system;

(d) Obsession with the solution stage;

(e) Assumption that the "model" is the real system;

(f) Communication in interdisciplinary efforts.

The first item in this list requires no further comment; the second includes both theoretical and computational difficulties in the mathematical treatment of a given set of equations. Although formidable obstacles can still arise, this is a much less critical problem today in modeling than it was, say, in the physical sciences in Newton's time. This is due in large part to great strides that have been made in the last several decades with the advance of modern computing facilities and the concomitant development of rather sophisticated numerical procedures. We remark that (c) is especially prevalent in certain physiological modeling, where systems are not easily manipulated experimentally. *In vitro* data and parameter values (determined via experimentation in nonphysiological ranges) are often used to model, predict and draw conclusions about *in vivo* situations. While (d) is likely to be a problem for investigators with a mathematical or physics background (in their enthusiasm for finding solutions of their model equations and various generalizations, they tend to forget or ignore the fact that the model is only an approximation and that certain aspects of the physical or biological model on which it is based are very poorly understood), item (e) can be a problem for both mathematical and physical and/or biological scientists. Even physicists and biologists sometimes have a penchant for disbelieving data that contradicts model simulations and predictions. It can be very tempting to throw out "faulty" data rather than reformulate the basic model. Finally, because most serious physical and biological modeling projects involve an interdisciplinary effort, there is always the possibility of serious lack of communication and cooperation due to differences in vocabulary, goals, and attitudes. Often mathematicians are only looking for a "problem" to which their already highly developed theories and techniques apply; i.e., they are in possession of a "solution" and in search of the "problem" they have solved! On the other hand, physicists and biologists can be too impatient with the mathematicians' desire to hypothesize rather implausible mechanisms and relationships (which can sometimes lead to exciting new perspectives about a phenomenon!).

Finally, we turn to the question of how one appraises a specific modeling attempt. There are a number of criteria that one might use. Among those proposed by various authors are the suggestions that a good model should: fit data accurately; be theoretically consistent with the real system; have parameters with physical meaning that can be measured independently of each other; prove useful in prediction; not so much explain or predict, but organize and economize thinking; pose new empirical questions and help answer them through the iterative process; help us understand the phenomena it represents and think comfortably about them; and point to inadequacies in some way of available data. It is clear, though, that for a modeling investigation to be deemed a success, it must have enhanced our overall knowledge and understanding of the phenomena in question. As one of our students (having been attacked by other students for some rather unorthodox and, at the time, unsupported hypothesis about mechanisms) noted in defending his efforts, "We learn little indeed if the models we build never stretch our understanding, but only tell us what we already feel is safely known." We remind the reader of the often quoted truth, "all models are incorrect, but some are more useful than others."

In concluding our philosophical remarks, we remark that one can distinguish between at least two basic types of scientific models: *descriptive* and *conceptual* models. Descriptive models, those designed to explain observed phenomena, will be the focus of our attention here. Conceptual models, models constructed to elucidate delicate and difficult points in some scientific theory, are often used to help resolve apparent paradoxes involving two descriptive models. Conceptual models do not appear widely in the biological literature since in many cases basic descriptive models are still under development.

Chapter 2

Modeling and Inverse Problems

In this chapter we will present a simple application to illustrate the iterative modeling process that was given in Chapter 1. In addition, using this illustrative example, the notion of an inverse problem will be discussed. The inverse or parameter estimation problem plays an indispensable role in developing mathematical models for biological and physical systems. As we shall see, because so many different mathematical models are plausible for a given system, *model validation* is an essential part of the modeling process. Indeed as we formulate the physical or biological problems mathematically, we find that the problem amounts to that of determining one or more unknown parameters in the mathematical model from some (limited) knowledge about the behavior of the system. Problems of this type arise in many important applications including geophysics, ecology, flexible structures, medical imaging and materials testing.

2.1 Mechanical Vibrations

To begin, we consider a spring of length l attached to a rigid horizontal support (e.g., ceiling) and a small object of mass m hanging from the bottom of the spring (see Figure 2.1). Now note that if we pull down or push up on the body a distance Δl, the elastic spring will exert a restoring force to pull the object back up or to push the object down, respectively. If Δl is small compared to the spring natural length l, then the spring restoring force, denoted by F_r, can be described by Hooke's law (see, e.g., [7, 11]). Mathematically, we write

$$F_r = -k\Delta l, \tag{2.1}$$

where k is called the spring-constant, which is a measure of the stiffness of the spring. Note that if Δl is *positive*, then the restoring force is *negative*, whereas if Δl is *negative*, then F_r is *positive*.

In modeling the motion of the mass m, it will be convenient to describe the position of the mass with respect to its equilibrium position. The equilibrium position of the mass is that point where the mass will hang at rest when no

FIGURE 2.1: Spring-mass system (with the mass in equilibrium position).

external forces (other than gravity) are being applied. We let $y = 0$ denote this equilibrium point and take the downward direction to be positive. Newton's Second Law of Motion is fundamental to the description of the position of the mass at time t; this states

$$F = ma, \tag{2.2}$$

where F is the sum of all forces exerted on the mass, m is the body's mass, and a is the acceleration of the body. Let $y(t)$ denote the position of the mass at time t. Using Newton's Second Law of Motion, we obtain (see for instance, [4, 5])

$$m\ddot{y} = -ky. \tag{2.3}$$

The differential equation (2.3) is a second-order, linear differential equation with constant coefficients. Its solution can be readily obtained as (see, e.g., [4, 10])

$$y(t) = A\cos\omega t + B\sin\omega t, \tag{2.4}$$

where $\omega = \sqrt{k/m}$. The constants of integration A and B are determined from the initial conditions, $y(0) = y_0$ and $\dot{y}(0) = v_0$, and are given by

$$A = y_0, \tag{2.5}$$

$$B = \frac{v_0}{\omega}. \tag{2.6}$$

In order to analyze the solution (2.4), it is convenient to rewrite it as a single cosine function of the form

$$y(t) = R\cos(\omega t - \phi), \tag{2.7}$$

where $R = \sqrt{A^2 + B^2}$ and $\phi = \tan^{-1}(B/A)$. This solution is depicted in Figure 2.2. Note that the solution $y(t)$ lies between $-R$ and $+R$, and that the motion of the body is periodic with a period of $2\pi/\omega$. This type of motion is called simple harmonic motion, $\omega = \sqrt{k/m}$ is called the natural frequency of the system, R is the amplitude of the motion, and ϕ is called the phase angle of the motion.

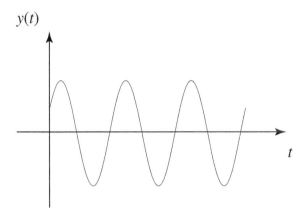

FIGURE 2.2: Graph of the simple harmonic motion, $y(t) = R\cos(\omega t - \phi)$.

In summary, our modeling process as discussed in Chapter 1 begins with the real physical model (a weight (mass) hanging from the bottom of an elastic spring) (step (i)) and proceeds with force balancing (Newton's Second Law of Motion) (step (ii)) to derive a mathematical model in terms of a differential equation (step (iii)). We next obtain the analytical solution to our differential equation model (step (v)). However, in comparison to real physical systems of mechanical vibration (step (vi)), the oscillations do not persist over time but eventually die out. This leads us to step (vii) which requires a re-examination in our understanding of the mechanical vibration system. Perhaps, we have over simplified our assumptions. For example, are there other forces (in addition to the spring restoring force) being exerted on the body in Newton's Second Law of Motion (2.2)? Specifically, consider a new experiment where we now add to the mass two light "massless" paddles (see Figure 2.3). As the body moves through the air, there is an apparent resistive force to motion (the paddles are bending in the direction opposite to motion). Furthermore, more bending will occur as the mass moves faster. Simply stated, this force is proportional to the magnitude of the velocity $\dot{y}$ and can be modeled by

$$F_d = -c\dot{y}, \tag{2.8}$$

where c is the viscous damping coefficient. The resistive force F_d, which the medium exerts on the body m, is also called the damping, or drag force. If we take this new force into consideration, our new mathematical model becomes

$$m\ddot{y} = -ky - c\dot{y}. \tag{2.9}$$

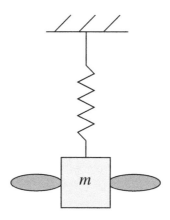

FIGURE 2.3: Spring-mass system (with "massless" paddles attached to the body).

If we assume that $c^2 - 4km < 0$, every solution of (2.9) has the form

$$y(t) = e^{-ct/2m}[A\cos\nu t + B\sin\nu t], \tag{2.10}$$

where $\nu = \frac{\sqrt{4km-c^2}}{2m}$ and A and B are constants to be determined from initial conditions as earlier. If we use similar arguments to those in the undamped case, the damped solution (2.10) can also be rewritten in the form

$$y(t) = Re^{-ct/2m}\cos(\nu t - \delta). \tag{2.11}$$

Observe that the solution oscillates between the curves $\pm Re^{-ct/2m}$. That is, the motion of the mass is periodic with decreasing amplitude, as depicted in Figure 2.4.

Thus, with damping present in the system, the motion of the body always dies out eventually. Engineers usually refer to such systems as spring-mass-dashpot systems. Spring-mass-dashpot systems are ubiquitous in engineering, science, and indeed in nature. For example, they are used as shock absorbers in vehicles to damp out bumps on the road as well as to minimize the recoil effect of a heavy gun barrel. They also find modeling applications in muscle mechanics and molecular level phenomena in materials (e.g., polarization, "electron cloud" models, in response to alternating electric fields).

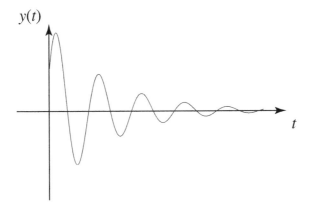

FIGURE 2.4: Plot of $y(t) = Re^{-ct/2m} \cos(\nu t - \delta)$.

2.2 Inverse Problems

Mathematical models as described by equations (2.3) and (2.9) above are of the "forward" type; that is, the parameters m, c, and k are assumed to be known, as well as the initial conditions. The mathematical model then predicts the resultant model behavior $y(t)$ at any time t from the solution formulas (2.4) or (2.10). This is typically the approach taken in sensitivity investigations, which is quite useful, and can provide important features of the model as functions of parameters (see [2, 6, 9, 12] and the references therein). However, in reality, not all parameters are directly measurable (e.g., most springs in mechanical devices come without specification of the spring constant k). Instead, we may have sparse and noisy measurements of displacements (using proximity sensors) and/or accelerations (using accelerometers). From this information, we need to find the unknown parameters. Problems of this type are called inverse or parameter estimation problems and are ubiquitous in modeling. Finding the solutions to an inverse problem is, in general, nontrivial because of non-uniqueness difficulties that arise. This undesirable feature is often due to noisy data and insufficient number of observations. For a discussion on the non-uniqueness as well as other issues such as stability in inverse problems we refer the interested reader to [1, 3].

To discuss the inverse problem formulation for the spring-mass-dashpot system, we assume that all three parameters m, c, and k are unknown and that displacement observations y_i^d at selected temporal points t_i are available. If we have noise free observations (which is never the case in practice), then we only need three well-chosen points t_i to obtain three equations to solve for three unknowns m, c, and k. However, due to noise in the measurements, we usually take n observations. Then, a typical inverse or estimation problem involving (2.9) is to find $q \in Q_{AD} = \{(m,c,k)|0 < m < M, 0 < c, 0 < k\}$ by minimizing the least squares criterion

$$J(q) = \sum_{i=1}^{n} |y_{\text{mod}}(t_i; m, c, k) - y_i^d|^2. \tag{2.12}$$

Here $y_{\text{mod}}(t_i; m, c, k)$ is the solution to (2.9) corresponding to m, c, and k. The above procedure leads to a constrained optimization problem. We also remark that such problems also require one to solve for the solution of the differential equation model (2.9) multiple times.

Project: Inverse Problem

The objective of this project is to help students familiarize themselves with the concepts involved with inverse problems. In addition, students will learn how to use MATLAB (see, e.g., [8]) to carry out many computations associated with inverse problems.

1.) Consider the following mathematical model for a spring-mass-dashpot system (using a linear spring assumption, Hooke's law, and viscous air damping):

$$m\frac{d^2y(t)}{dt^2} + c\frac{dy(t)}{dt} + ky(t) = 0$$

with initial conditions

$$y(0) = 2, \qquad \frac{dy(0)}{dt} = 0,$$

where m is the mass, c is the damping coefficient, k is the spring constant, and $y(t)$ is the vertical displacement of the mass from the equilibrium position.

The solution to the above second order differential equation can be computed using MATLAB routine ode23. To use the routine ode23 one needs to rewrite the above equation as a system of first order differential equations. That is, letting $z_1 = y$ and $z_2 = \frac{dy}{dt}$, we obtain

$$\frac{dz_1}{dt} = z_2$$
$$\frac{dz_2}{dt} = -\frac{k}{m}z_1 - \frac{c}{m}z_2.$$

Letting $m = 2$, $c = 2$, $k = 3$, compute the numerical solution $y(t)$ for $t \in [0,5]$ and plot it on a graph. On your graph, you should label the horizontal axis as *time, t*, the vertical axis as *y(t)*, and title the graph as *A Linear Spring Model Response*. Also, place a text string on the graph showing values of m, c, and k. The following MATLAB functions will be useful for this exercise: `ode23`, `plot`, `xlabel`, `ylabel`, `title` and `gtext`. You can see an explanation of a function by typing `help ode23`, for example, on the command line.

2.) In general, the coefficients m, c, and k are unknown parameters. These parameters can be estimated via a nonlinear least squares estimation problem. Specifically one seeks $\vec{q} = (m, c, k)$ to minimize the cost function

$$J(\vec{q}) = \sum_{i=1}^{n} \left| y_m(t_i; \vec{q}) - y_i^d \right|^2$$

where $y_m(t_i; \vec{q})$ is the model solution to the spring-mass-dashpot model at time t_i for $i = 1, 2, \ldots, n$, given the parameter set $\vec{q}$ and y_i^d is the data (displacement) collected also at time t_i. In this exercise, we will create "*simulated*" data to be used for estimating the unknown parameters $\vec{q} = (m, c, k)$. For this, we assume that displacement is sampled at equally spaced time intervals. We will subdivide the time interval $[0, 5]$ into n equal subintervals of length $h = 5/n$. Let y_i^d denote the displacement sampled at time $t_i = ih$, $i = 1, \ldots, N$. For this, use the solution $y(t_i)$ to the spring-mass-dashpot system corresponding to $m = 2$, $c = 2$, and $k = 3$ that you have already computed in part 1.) of this exercise. Using these "data", <u>implement</u> an inverse problem for finding the parameters m, c, and k <u>using the least</u> squares criteria above (take $n = 20$). The solution to this minimization problem can be solved using MATLAB routine `fminu` or `fminsearch`. To use one of these routines you must give an initial guess for the parameters (<u>try</u> $\vec{q}_g = (m, c, k) = (3, 1, 6)$; <u>then</u> try several others). Create a table showing the initial guess values of the parameters $\vec{q}_g$, its cost function value $J(\vec{q}_g)$ and the optimal values of the parameters $\vec{q}_{op}$ and its cost function value $J(\vec{q}_{op})$.

3.) In practice, the collected data is corrupted by noise (for example, errors in collecting data, instrumental errors, etc.). In the next part of the exercise, we wish to test the sensitivity of the inverse least squares method to errors in sampling the data. For this, we will add to each simulated data point an error term as follows:

$$\hat{y}_d(t_i) = y_d(t_i) + nl \cdot rand_i,$$

where $rand_i$ are the normally distributed random numbers with zero mean and variance 1.0. Use the MATLAB routine `randn` to <u>generate</u> an n-vector with random entries. Here, nl is a noise level constant.

For each of the values $nl = 0.01$, $nl = 0.02$, $nl = 0.05$, $nl = 0.1$, $nl = 0.2$, <u>estimate</u> the parameters m, c, and k using the inverse least squares method. <u>Create</u> a table listing the estimated values of the parameters and the values of the cost functionals for each value of nl. <u>Describe</u> the sensitivity of the inverse least squares method with respect to the noise level nl.

References

[1] R. Aster, B. Borchers and C. Thurber, *Parameter Estimation and Inverse Problems*, Academic Press, New York, 2004.

[2] H.T. Banks, S. Dediu and S.E. Ernstberger, Sensitivity functions and their uses in inverse problems, *J. Inverse and Ill-posed Problems*, **15**, 2007, pp. 683–708.

[3] H.T. Banks and K. Kunisch, *Estimation Techniques for Distributed Parameter Systems*, Birkhäuser, Boston, 1989.

[4] W.E. Boyce and R.C. DiPrima, *Elementary Differential Equations and Boundary Value Problems*, John Wiley & Sons, Inc., Hoboken, 8th ed., 2004.

[5] M. Braun, *Differential Equations and Their Applications: An Introduction to Applied Mathematics*, Springer, Berlin, 4th ed., 1992.

[6] J.A. David, *Optimal Control, Estimation, and Shape Design: Analysis and Applications*, Ph.D. Dissertation, North Carolina State University, Raleigh, 2007.

[7] J.M. Gere and S.P. Timoshenko, *Mechanics of Materials*, PWS Pub. Co., Boston, 4th ed., 1997.

[8] A. Gilat, *MATLAB: An Introduction with Applications*, John Wiley & Sons, Inc., Hoboken, 2nd ed., 2004.

[9] E. Laporte and P. Le Tallec, *Numerical Methods in Sensitivity Analysis and Shape Optimization*, Birkhäuser, Boston, 2002.

[10] R.K. Nagle, E.B. Saff and A.D. Snider, *Fundamentals of Differential Equations and Boundary Value Problems*, Pearson Education, Inc., Boston, 2004.

[11] S.S. Rao, *Vibration of Continuous Systems*, John Wiley & Sons, Inc., Hoboken, 2007.

[12] A. Saltelli, K. Chan and E.M. Scott, *Sensitivity Analysis*, John Wiley & Sons, Inc., Hoboken, 2000.

Chapter 3

Mathematical and Statistical Aspects of Inverse Problems

In inverse or parameter estimation problems as discussed in Chapter 2, an important but practical question is how successful the mathematical model is in describing the physical or biological phenomena represented by the experimental data. In general, it is very unlikely that the residual sum of squares (RSS) in the inverse least squares formulation is zero. Indeed, due to modeling error, there may not even be a true set of parameters so that the mathematical model will provide an exact fit to the experimental data.

Even if one begins with a deterministic model and has no initial interest in uncertainty or stochasticity, as soon as one employs experimental data in the investigation, one is led to uncertainty that should not be ignored. This is because all measurement procedures contain error or uncertainty in the data collection process and hence statistical questions arise. To correctly formulate, implement and analyze the corresponding inverse problems one requires a framework entailing a *statistical model* as well as a *mathematical model*.

In this chapter we discuss mathematical, statistical and computational aspects of inverse or parameter estimation problems for deterministic dynamical systems based on the Maximum Likelihood Estimation (MLE), Ordinary Least Squares (OLS) and Generalized Least Squares methods (GLS) with appropriate corresponding data noise assumptions of constant variance and nonconstant variance (relative error), respectively, in the latter two cases. Among the topics included here are interplay between the mathematical model, the statistical model and observation or data assumptions, and some techniques (residual plots and model comparison tests) for analyzing the uncertainties associated with inverse problems employing experimental data. We also outline a standard theory underlying the construction of confidence intervals for parameter estimators. The methodology for statistical "hypothesis" testing that can form the basis of a heuristic approach to address the important problem of model improvement is illustrated. This latter approach along with a number of examples as well as its mathematical theory can be found in the monograph by Banks and Kunisch [8] while the asymptotic theory for confidence intervals can be found in Seber and Wild [17]. A recent summary [2] contains more examples along with extensive references.

Before we begin the inverse problem discussions, we give a brief but useful review of certain basic probability and statistics concepts.

3.1 Probability and Statistics Overview

The theory of probability and statistics is an essential mathematical tool in the development of inverse problem formulations and subsequent analysis as well as for approaches to statistical hypothesis testing. Our coverage of these fundamental and important topics is brief and limited in scope. Indeed, we provide in this section a few definitions and basic concepts in the theory of probability and statistics that are essential for the understanding of estimators, confidence intervals and hypothesis testing to be formulated later in the chapter.

3.1.1 Probability

We adopt the standard practice of denoting events by capital letters and will write the probability of event A as $P(A)$. The set of all possible outcomes, the *sample space*, will be denoted by S. For example, consider the experiment of the rolling of a die in which there are six possible outcomes. The sample space is

$$S = \{1, 2, 3, 4, 5, 6\} \tag{3.1}$$

and an event A might be defined as

$$A = \{1, 5\}, \tag{3.2}$$

which consists of the outcomes 1 and 5. Associated with event A contained in S is its probability $P(A)$. In the case that the sample space S is discrete (finite or countably infinite) probability satisfies the following postulates:

(i) $P(A) \geq 0$,

(ii) $P(S) = 1$,

(iii) If $A_1, A_2, A_3, \ldots$, is a finite or an infinite sequence of disjoint subsets of S, then

$$P(A_1 \cup A_2 \cup A_3 \cup \cdots) = P(A_1) + P(A_2) + P(A_3) + \cdots . \tag{3.3}$$

For example, in our fair experiment of the rolling of a die, each possible outcome has probability $\frac{1}{6}$. The event A as defined by (3.2) consists of two disjoint subevents, and hence $P(A) = \frac{2}{6} = \frac{1}{3}$. Using the three postulates of probability, a number of immediate consequences can also be derived which have important applications. For example, probabilities cannot exceed 1 ($P(A) \leq 1$ for any event A), the empty set $\emptyset$ has probability 0 ($P(\emptyset) = 0$), and the probability that an event will occur and that it will not occur always add up to 1 ($P(A) + P(\bar{A}) = 1$ where $\bar{A}$ denotes the complement of the event A which consists of all sample points in S that are not in A). If $P(A) = 1$, then we say that "the event A occurs with probability 1 or *almost surely* (a.s.)."

Instead of considering a single experiment, let us perform two experiments and consider their outcomes. For example, the two experiments may be two separate tosses of a single die or a single toss of two dice. The sample space in this case consists of 36 pairs (i, j), where $i, j = 1, 2, \ldots, 6$. Note that in a fair dice game, each point in the sample space has probability $\frac{1}{36}$. We now consider the probability of joint events, such as $\{i = 2, j = \text{odd}\}$. We begin by denoting the possible outcomes of one experiment by A_i, $i = 1, 2, \ldots, n$, and by B_j, $j = 1, 2, \ldots, m$ the possible outcomes of the second experiment. The combined experiment has the possible joint outcomes (A_i, B_j), where $i = 1, 2, \ldots, n$ and $j = 1, 2, \ldots, m$. The joint probability $P(A_i, B_j)$ satisfies the condition

$$0 \leq P(A_i, B_j) \leq 1. \tag{3.4}$$

If the outcomes B_j for $j = 1, 2, \ldots, m$ are mutually exclusive (i.e., $B_i \cap B_j = \emptyset, i \neq j$), then

$$\sum_{j=1}^{m} P(A_i, B_j) = P(A_i). \tag{3.5}$$

Furthermore, if all the outcomes of the two experiments are mutually exclusive, then

$$\sum_{i=1}^{n} \sum_{j=1}^{m} P(A_i, B_j) = 1. \tag{3.6}$$

The generalization of the above concept to more than two experiments follows in a straightforward manner.

Next, we consider a joint event with probability $P(A, B)$. Assuming that event A has occurred, we wish to determine the probability of the event B. This is called the *conditional probability* of the event B given the occurrence of the event A and is given by

$$P(B|A) = \frac{P(A, B)}{P(A)}, \tag{3.7}$$

where $P(A) > 0$. A very useful relationship for conditional probabilities, which is known as Bayes' theorem, states that if A_i, where $i = 1, 2, \ldots, n$, are mutually exclusive events such that

$$\bigcup_{i=1}^{n} A_i = S \tag{3.8}$$

and B is an arbitrary event with $P(B) > 0$, then

$$P(A_i|B) = \frac{P(A_i, B)}{P(B)}$$

$$= \frac{P(B|A_i)P(A_i)}{\sum_{l=1}^{n} P(B|A_l)P(A_l)}.$$

3.1.2   Random Variables

In most applications of probability theory, we are interested only in a particular aspect of the outcome of an experiment. For example, in the experiment of the rolling of a pair of dice, we are generally interested only in the total and not in the outcome for each die. In the language of probability and statistics, the total which we obtain with a pair of dice is called a *random variable*. More formally, the random variable $X(A)$ represents the functional relationship between a random event A and a real number. For example, if we flip a coin the possible outcomes are heads, H, or tails, T. We may define a random variable $X(A)$ by

$$X(A) = \begin{cases} 1, & A = H, \\ -1, & A = T. \end{cases} \tag{3.9}$$

We note that the random variable may be continuous or discrete. Associated with a random variable X, we consider the event $X \leq x$, where $-\infty < x < \infty$. The probability of this event is defined by

$$F(x) = P(X \leq x), \tag{3.10}$$

where the function $F(x)$ is called the *probability distribution function* of the random variable X. It is also called the *cumulative distribution function* or *(cdf)*. The distribution function is right continuous and has the following properties:

(i) $0 \leq F(x) \leq 1$,

(ii) $F(x_1) \leq F(x_2)$ if $x_1 \leq x_2$,

(iii) $F(-\infty) = 0$,

(iv) $F(\infty) = 1$.

The derivative $p(x)$ (when it exists) of the distribution function $F(x)$ given by

$$p(x) = \frac{dF(x)}{dx} \tag{3.11}$$

is called the *probability density function* or *(pdf)*. The name "density function" comes from the fact that the probability of the event $x_1 \leq X \leq x_2$ is given by

$$P(x_1 \leq X \leq x_2) = P(X \leq x_2) - P(X \leq x_1)$$
$$= F(x_2) - F(x_1)$$
$$= \int_{x_1}^{x_2} p(x)\, dx.$$

The probability density function $p(x)$ satisfies the following properties:

(i) $p(x) \geq 0$,

(ii) $\int_{-\infty}^{\infty} p(x)\, dx = F(\infty) - F(-\infty) = 1$.

Moreover, it is common to denote *random variables* by capital letters X, Y, Z, etc., while one denotes particular *realizations* by the corresponding lower case letters x, y, z, etc.

3.1.3 Statistical Averages of Random Variables

Of particular importance in the characterization of the outcomes of experiments and random variables are the concepts of first and second moments of a single random variable and the joint moments (correlation and covariance) between any pair of random variables in a multi-dimensional set of random variables.

We begin the discussion of these statistical averages by considering first a single random variable X and its pdf $p(x)$. The *mean value* μ or *expected value* of the random variable X is defined by

$$\mu = E(X) = \int_{-\infty}^{\infty} x p(x)\, dx, \tag{3.12}$$

where $E(\cdot)$ is called the expected value operator (or statistical averaging operator). This is the *first moment* of the random variable X. The nth moment of a probability distribution of a random variable X is defined as

$$E(X^n) = \int_{-\infty}^{\infty} x^n p(x)\, dx. \tag{3.13}$$

We can also define the *central moments*, which are the moments of the difference between X and μ. The second central moment, which is called the *variance* of X, is defined by

$$\sigma^2 = \text{var}(X) = E[(X - \mu)^2] = \int_{-\infty}^{\infty} (x - \mu)^2 p(x)\, dx. \tag{3.14}$$

The square root σ of the variance of X is called the *standard deviation* of X. Variance is a measure of the "randomness" of the random variable X. It is related to the first and second moments through the relationship

$$
\begin{aligned}
\sigma^2 &= E(X^2 - 2\mu X + \mu^2) \\
&= E(X^2) - 2\mu E(X) + \mu^2 \\
&= E(X^2) - \mu^2.
\end{aligned}
$$

In the important case of multi-dimensional or R^p–valued vector random variables $X = (X_1, X_2, \ldots, X_p)^T$, we can define joint moments of any order. However, the joint moments that are most useful in practical applications are the joint moments defined by

$$
E(X_i X_j) = \int_{-\infty}^{\infty} \int_{-\infty}^{\infty} x_i x_j p(x_i, x_j)\, dx_i dx_j, \tag{3.15}
$$

which are called the *correlation* (not to be confused with *correlation coefficients*) between the random variables X_i and X_j. Here, $p(x_i, x_j)$ are the *marginal densities* defined by

$$
\begin{aligned}
&p(x_i, x_j) \\
&= \int_{-\infty}^{\infty} \cdots \int_{-\infty}^{\infty} p(x_1, \ldots, x_p) dx_1 \ldots dx_{i-1} dx_{i+1} \ldots dx_{j-1} dx_{j+1} \ldots dx_p.
\end{aligned}
$$

Also of particular importance is the joint central moment, which is also called the *covariance* of X_i and X_j and is given by

$$
\begin{aligned}
\mu_{ij} &\equiv E[(X_i - \mu_i)(X_j - \mu_j)] \\
&= \int_{-\infty}^{\infty} \int_{-\infty}^{\infty} (x_i - \mu_i)(x_j - \mu_j) p(x_i, x_j)\, dx_i dx_j \\
&= \int_{-\infty}^{\infty} \int_{-\infty}^{\infty} x_i x_j p(x_i, x_j)\, dx_i dx_j - \mu_i \mu_j \\
&= E(X_i X_j) - \mu_i \mu_j. \tag{3.16}
\end{aligned}
$$

The $(p \times p)$ matrix with elements μ_{ij} is called the *covariance matrix* of the random variable $X = (X_1, \ldots, X_p)$. Two random variables are said to be uncorrelated if $E(X_i X_j) = E(X_i)E(X_j) = \mu_i \mu_j$. In that case, the covariance $\mu_{ij} = 0$. We also note that when X_i and X_j are statistically independent, they are uncorrelated. The reverse is, however, not true. That is, if X_i and X_j are uncorrelated, they are not necessarily statistically independent.

3.1.4　Special Probability Distributions

We say a continuous random variable has a certain *distribution* (e.g., *Gaussian distribution*) when it has the corresponding probability density. We review in this section several frequently encountered random variables, their pdf's, and their moments.

Uniform distribution. The pdf of a uniformly distributed random variable X is given by

$$p(x) = \begin{cases} 1/(b-a), & a \le x \le b, \\ 0, & \text{otherwise.} \end{cases} \qquad (3.17)$$

This is also called a rectangular distribution and its graph is depicted in Figure 3.1. The first two moments of X are

$$E(X) = \mu = \frac{a+b}{2}$$

$$E(X^2) = \frac{a^2 + b^2 + ab}{3}$$

and the variance is

$$\text{var}(X) = \sigma^2 = \frac{(a-b)^2}{12}. \qquad (3.18)$$

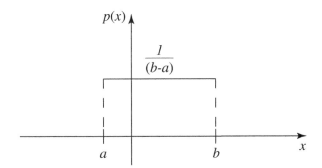

FIGURE 3.1: Plot of the pdf $p(x)$ of a uniform distribution.

Gaussian (normal) distribution. The pdf of a Gaussian or normally distributed random variable is given by

$$p(x) = \frac{1}{\sqrt{2\pi}\sigma} e^{-(x-\mu)^2/(2\sigma^2)}, \qquad (3.19)$$

where μ is the mean and σ^2 is the variance of the random variable. The pdf of a Gaussian distributed random variable is illustrated in Figure 3.2. The probability distribution function $F(x)$ has the form

$$F(x) = \int_{-\infty}^{x} p(s)\, ds$$

$$= \frac{1}{\sqrt{2\pi}\sigma} \int_{-\infty}^{x} e^{-(s-\mu)^2/(2\sigma^2)}\, ds$$

$$= \frac{1}{2}\left[1 + \mathrm{erf}(\frac{x-\mu}{\sqrt{2}\sigma})\right], \tag{3.20}$$

where erf denotes the error function and is given by

$$\mathrm{erf}(x) = \frac{2}{\pi} \int_{0}^{x} e^{-s^2}\, ds. \tag{3.21}$$

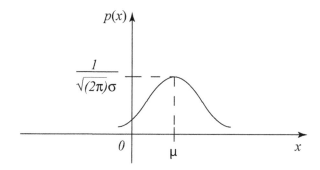

FIGURE 3.2: The pdf graph of a Gaussian distributed random variable.

The kth central moment of the random variable X is given by the expression

$$E[(X-\mu)^k] \equiv m_k = \begin{cases} 1\cdot 3 \cdots (k-1)\sigma^k, & k = \text{even}, \\ 0, & k = \text{odd} \end{cases} \tag{3.22}$$

and the kth moments are given in terms of the central moments by

$$E(X^k) = \sum_{i=0}^{k} \binom{k}{i} \mu^i m_{k-i}. \tag{3.23}$$

Finally, if X is a random variable distributed normally with mean μ and variance σ^2, this is commonly denoted by $X \sim \mathcal{N}(\mu, \sigma^2)$.

Log-normal distribution. A widely employed model for biological (and other) phenomena where the random variable is only allowed positive values is the so-called *log-normal* distribution. If $\log X \sim \mathcal{N}(\mu, \sigma^2)$, then X has a log-normal distribution with density given by

$$p(x) = \frac{1}{\sqrt{2\pi}\sigma} \frac{1}{x} \exp\left\{-\frac{(\log x - \mu)^2}{2\sigma^2}\right\}, \quad 0 < x < \infty, \tag{3.24}$$

with mean and variance

$$E(X) = e^{\mu + \sigma^2/2}, \tag{3.25}$$

$$\operatorname{var}(X) = (e^{\sigma^2} - 1)e^{2\mu + \sigma^2}. \tag{3.26}$$

We observe that $\operatorname{var}(X)$ is proportional to $\{E(X)\}^2$ so that the constant *coefficient of variation (CV)* defined by $\sqrt{\operatorname{var}(X)}/E(X)$, which represents the "*noise-to-signal*" ratio, does not depend on $E(X)$. The density for this random variable is skewed (asymmetric) with a "*long right tail*" but becomes more and more symmetric as $\sigma \to 0$.

Multivariate normal distribution. One of the most often encountered multivariate random variables is (as we shall see below in discussing asymptotic theory for confidence intervals) also incredibly important in statistical *modeling* and *inference* and is known as the *multivariate normal* or *multinormal* random variable. A *random vector* $X = (X_1, \ldots, X_p)^T$ has a multivariate (p-variate) normal distribution (denoted by $X \sim \mathcal{N}_p(\mu, \Sigma)$) if $\alpha^T X$ is normal for all $\alpha \in R^p$; its density is given by

$$p(x) = (2\pi)^{-p/2} |\Sigma|^{-1/2} \exp\{-(x - \mu)^T \Sigma^{-1}(x - \mu)/2\},$$

for $x = (x_1, \ldots, x_p)^T \in R^p$ where the mean is

$$\mu = E(X) = (\mu_1, \ldots, \mu_p)^T = \{E(X_1), \ldots, E(X_p)\}^T$$

and the *covariance matrix* is

$$\Sigma = E\{(x - \mu)(x - \mu)^T\}.$$

The ($p \times p$) covariance matrix Σ is such that

$$\Sigma_{jj} = \operatorname{var}(X_j), \quad \Sigma_{jk} = \Sigma_{kj} = \operatorname{cov}(X_j, X_k).$$

Finally, we note that the *marginal* probability densities are *univariate* normal.

Chi-square distribution. The chi-square distribution is important in statistical analysis of variance (ANOVA) and other statistical procedures [10, 14] based on normally distributed random variables. In particular, a chi-square distributed random variable is related to the normally distributed random variable through a transformation. That is, if X is a normally distributed random variable, then $Y = X^2$ has a chi-square distribution. There are two types of chi-square distributions. A central chi-square distribution is obtained when X has zero mean; otherwise, we call it a non-central chi-square distribution.

First, let us consider the central chi-square distribution. In this case, the pdf of Y has the form

$$p(y) = \frac{1}{\sqrt{2\pi y}\sigma} e^{-y/(2\sigma^2)}, \tag{3.27}$$

where $y > 0$. The corresponding probability distribution function $F(y)$ is given by

$$F(y) = \frac{1}{\sqrt{2\pi}\sigma} \int_0^y \frac{1}{\sqrt{s}} e^{-s/(2\sigma^2)} \, ds. \tag{3.28}$$

More generally, suppose that the random variable Y is defined as

$$Y = \sum_{i=1}^{k} X_i^2, \tag{3.29}$$

where X_i, $i = 1, 2, \ldots, k$, are statistically independent and identically distributed normal random variables with zero mean and variance σ^2. The pdf is then given by

$$p(y) = \frac{1}{\sigma^k 2^{k/2} \Gamma(k/2)} y^{k/2-1} e^{-y/(2\sigma^2)}, \qquad y > 0, \tag{3.30}$$

where $\Gamma(p)$ is the gamma function defined as $\Gamma(1/2) = \sqrt{\pi}$, $\Gamma(3/2) = \sqrt{\pi}/2$, and for $\nu > 0$,

$$\Gamma(\nu) = \int_0^\infty x^{\nu-1} e^{-x} \, dx. \tag{3.31}$$

By integration by parts, it can be shown that

$$\Gamma(\nu) = (\nu - 1)\Gamma(\nu - 1). \tag{3.32}$$

For positive and integer ν we obtain

$$\Gamma(\nu) = (\nu - 1)!. \tag{3.33}$$

This pdf, which is a generalization of (3.27), is called a *chi-square (or gamma) pdf with k degrees of freedom* (denoted $Y \sim \chi^2(k)$ or $Y \sim \chi_k^2$). Its graphs for several values of k are depicted in Figure 3.3. The first two moments of Y are

$$E(Y) = k\sigma^2$$
$$E(Y^2) = 2k\sigma^4 + k^2\sigma^4$$

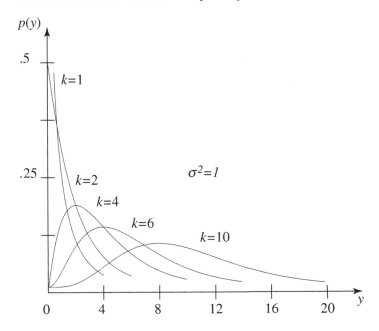

FIGURE 3.3: The pdf graph of a chi-square distribution for various degrees of freedom k.

and its variance is

$$\text{var}(Y) = 2k\sigma^4. \tag{3.34}$$

We now turn to the non-central chi-square distribution. Here, let X_i, $i = 1, 2, \ldots, k$, be Gaussian distributed random variables with means μ_i and identical variances equal to σ^2. The random variable $Y = \sum_i^k X_i^2$ has the pdf

$$p(y) = \frac{1}{2\sigma^2} \left(\frac{y}{s}\right)^{(k-2)/4} e^{-(s^2+y)/(2\sigma^2)} I_{k/2-1}\left(\frac{\sqrt{y}s}{\sigma^2}\right), \qquad y > 0, \tag{3.35}$$

where the parameter s^2, which is called the noncentrality parameter of the distribution, is given by

$$s^2 = \sum_{i=1}^{k} \mu_i^2.$$

The function $I_\alpha(x)$ is the αth-order modified Bessel function of the first kind and is given by

$$I_\alpha(x) = \sum_{j=0}^{\infty} \frac{(x/2)^{\alpha+2j}}{j!\,\Gamma(\alpha + j + 1)}, \qquad x \geq 0. \tag{3.36}$$

The pdf function given by the expression (3.35) is called the *non-central chi-square pdf with k degrees of freedom*.

Finally, the first two moments of the non-central chi-square distribution random variable are

$$E(Y) = k\sigma^2 + s^2$$
$$E(Y^2) = 2k\sigma^4 + 4\sigma^2 s^2 + (k\sigma^2 + s^2)^2$$

and its variance is

$$\text{var}(Y) = 2k\sigma^4 + 4\sigma^2 s^2.$$

Rayleigh distribution. Another frequently encountered random variable which is closely related to the central chi-square distribution is the Rayleigh distribution. To begin the discussion, let us consider a central chi-square distribution with two degrees of freedom, $Y = X_1^2 + X_2^2$, where X_i are zero mean statistically independent Gaussian random variables with identical variances σ^2. The pdf of Y is given by

$$p(y) = \frac{1}{2\sigma^2} e^{-y/(2\sigma^2)}. \tag{3.37}$$

Define a new variable Z as

$$Z = \sqrt{X_1^2 + X_2^2} = \sqrt{Y}.$$

Then, after a change of variables in equation (3.37), we obtain the pdf of Z as

$$p(z) = \frac{z}{\sigma^2} e^{-z^2/(2\sigma^2)}, \qquad z \geq 0$$

which is known as the pdf of a Rayleigh distributed random variable. The corresponding probability distribution function is given by

$$F(z) = \int_0^z \frac{s}{\sigma^2} e^{-s^2/(2\sigma^2)} \, ds$$
$$= 1 - e^{-z^2/(2\sigma^2)}, \qquad z \geq 0.$$

The moments of Z are

$$E(Z^k) = (2\sigma^2)^{(k/2)} \Gamma(1 + \frac{k}{2})$$

and the variance is given by

$$\text{var}(Z) = (2 - \frac{\pi}{2})\sigma^2.$$

Student's t distribution. If $U \sim \mathcal{N}(0, 1)$ and $V \sim \chi^2(k)$ are independent, then $X = U/\sqrt{V/k}$ has a t distribution with k degrees of freedom (denoted by $X \sim t^k$) and density function

$$p(x) = \frac{\Gamma\{(k+1)/2\}}{\Gamma(k/2)} \frac{1}{\sqrt{k\pi}} \frac{1}{(1 + x^2/k)^{(k+1)/2}}, \quad -\infty < x < \infty.$$

The mean and variance are given by

$$E(X) = 0 \quad \text{if} \quad k > 1 \text{ (otherwise undefined)}$$

and

$$\text{var}(X) = k/(k-2) \quad \text{if} \quad k > 2 \text{ (otherwise undefined)}.$$

The corresponding density is symmetric like that of the normal, with "*heavier tails*," and becomes similar to a normal as $k \to \infty$. As we shall see below, the Student's t distribution is fundamental to the computation of confidence intervals for estimated parameters using experimental data in inverse problems (where typically $k \gg 2$).

3.2   Parameter Estimation or Inverse Problems

3.2.1   The Mathematical Model

We consider inverse or parameter estimation problems in the context of a parameterized (with vector parameter $\vec{q}$) dynamical system or **mathematical model**

$$\frac{d\vec{z}}{dt}(t) = \vec{g}(t, \vec{z}(t), \vec{q}) \tag{3.38}$$

with **observation process**

$$\vec{y}(t) = \mathcal{C}\vec{z}(t; \vec{q}). \tag{3.39}$$

The mathematical model is a deterministic system (here we treat ordinary differential equations, but our discussions are relevant to problems involving parameter dependent partial differential equations, delay differential equations, etc., as long as the system is assumed to be well-posed, i.e., to possess unique solutions that depend smoothly on the parameters and initial data). Following usual convention (which corresponds to the form of data usually available from experiments), we assume a discrete form of the observations in which one has n longitudinal observations $\vec{y}_j$ corresponding to

$$\vec{y}(t_j) = \mathcal{C}\vec{z}(t_j; \vec{q}), \quad j = 1, \ldots, n, \tag{3.40}$$

where $\mathcal{C}$ is an observation operator that will be described below. In general the corresponding observations or data $\{\vec{y}_j\}$ will not be exactly $\vec{y}(t_j)$. Because of the nature of the phenomena leading to this discrepancy, we treat this uncertainty pertaining to the observations with a statistical model for the observation process.

3.2.2 The Statistical Model

In our discussions here we consider a **statistical model** of the form

$$\vec{Y}_j = \vec{f}(t_j, \vec{q}_0) + \vec{\mathcal{E}}_j, \quad j = 1, \ldots, n, \tag{3.41}$$

where $\vec{f}(t_j, \vec{q}) = \mathcal{C}\vec{z}(t_j; \vec{q})$, $j = 1, \ldots, n$, corresponds to the observed part of the solution of the mathematical model (3.38) at the jth covariate or observation time for a particular vector of parameters $\vec{q} \in R^p, \vec{z} \in R^N, \vec{f} \in R^m$, and $\mathcal{C}$ is an $m \times N$ matrix. The term $\vec{q}_0$ represents the "truth" or the parameters that generate the observations $\{\vec{Y}_j\}_{j=1}^n$. (The existence of a truth parameter $\vec{q}_0$ is a standard assumption in statistical formulations and this along with the assumption that the means $E[\vec{\mathcal{E}}_j]$ are zero yields implicitly that (3.38) is a correct description of the process being modeled.) The terms $\vec{\mathcal{E}}_j$ are random variables which can represent observation or measurement error, "system fluctuations" or other phenomena that cause observations to not fall exactly on the points $\vec{f}(t_j, \vec{q})$ from the smooth path $\vec{f}(t, \vec{q})$. Since these fluctuations are unknown to the modeler, we will assume that realizations $\vec{\epsilon}_j$ of $\vec{\mathcal{E}}_j$ are generated from a probability distribution (with mean zero throughout our discussions) that reflects the assumptions regarding these phenomena. Thus specific data (*realizations*) corresponding to (3.41) will be represented by

$$\vec{y}_j = \vec{f}(t_j, \vec{q}_0) + \vec{\epsilon}_j, \quad j = 1, \ldots, n. \tag{3.42}$$

Assumptions about the distribution for $\vec{\mathcal{E}}_j$ must be problem specific. For instance, in a statistical model for pharmacokinetics of drugs in human blood samples, a natural distribution for $\vec{\mathcal{E}} = (\mathcal{E}_1, \ldots, \mathcal{E}_n)^T$ might be a multivariate normal distribution. In other applications the distribution for $\vec{\mathcal{E}}$ might be much more complicated [11]. For example, in observing (counting) populations, the error may well depend on the size of the population itself (i.e., so-called *relative error* to be discussed below).

To relate the notation and formulations of this chapter to that of the inverse problems introduced in the previous chapter, we observe that $y_j = y_j^d$ is the data and $f(t_j, \vec{q}) = y_{mod}(t_j; m, c, k) = y_m(t_j; m, c, k)$ in (2.12). Moreover, in the computational project of Chapter 2, $\vec{q} = (m, c, k)$, $\vec{z} = (y, \dot{y})^T$, so that $f(t_j, \vec{q}) = \mathcal{C}\vec{z}$ where $\mathcal{C} = (1\ 0)$.

The purpose of our presentation is to discuss methodology related to the estimation of the true value of the parameters $\vec{q}_0$ from a set $\mathcal{Q}$ of admissible parameters, and its dependence on what is assumed about the variance of the error $\vec{\mathcal{E}}_j$, $\mathrm{var}(\vec{\mathcal{E}}_j)$. We discuss three inverse problem methodologies that can be used to calculate estimates $\hat{q}$ for $\vec{q}_0$: the ordinary least-squares (OLS) and the generalized least-squares (GLS) formulations as well as the popular maximum likelihood estimate (MLE) formulation in the case where one assumes the distributions of the error process $\{\vec{\mathcal{E}}_j\}$ are known.

3.2.3 Known Error Processes: Maximum Likelihood Estimators

In the introduction of the statistical model we initially made no mention of the probability distribution that generates the error realizations $\vec{\epsilon}_j$. In many situations one readily assumes that the errors $\vec{\mathcal{E}}_j$, $j = 1, \ldots, n$, are independent and identically distributed (we make the *standing assumptions of independence across j* throughout our discussions in this chapter). We discuss a case where one is able to make further assumptions on the error, namely that the <u>distribution is known</u>. In this case, maximum likelihood techniques may be used. We discuss first one such case for a scalar observation system, i.e., $m = 1$. If $\mathcal{E}_j$ is assumed a known random variable with parameter $(\vec{\theta})$ dependent density $p_{\vec{\theta}}(\epsilon) = p(\epsilon; \vec{\theta})$, then for the statistical model (3.41) with observations $\vec{Y}$, the associated *likelihood function* is defined by

$$L(\vec{\theta}|\vec{Y}) = \prod_{j=1}^{n} p(Y_j - f(t_j; \vec{q}); \vec{\theta}). \tag{3.43}$$

In particular, one often assumes that $\vec{\theta} = (\mu, \sigma)$ and $p = p(\epsilon; \mu, \sigma^2)$ so that the density is completely characterized by its mean and variance. If we further assume that $\mathcal{E}_j$ has known density $p(\epsilon; 0, \sigma_0^2)$, then from the statistical model (3.41) we have that $Y_j - f(t_j, \vec{q})$ has density $p(\epsilon; 0, \sigma_0^2)$ or Y_j has the density $p(y; f(t_j, \vec{q}), \sigma_0^2)$. The corresponding likelihood function is

$$L(\vec{q}, \sigma^2|\vec{Y}) = \prod_{j=1}^{n} p(Y_j - f(t_j; \vec{q}); 0, \sigma^2). \tag{3.44}$$

3.2.3.1 Normally Distributed Errors

If, in addition, there is sufficient evidence to suspect that the error is generated by a normal distribution, then we may be willing to assume $\mathcal{E}_j \sim \mathcal{N}(0, \sigma_0^2)$, and hence $Y_j \sim \mathcal{N}(f(t_j, \vec{q}_0), \sigma_0^2)$. We can then obtain an expression for determining $\vec{q}_0$ and σ_0 by seeking the maximum over $(\vec{q}, \sigma^2) \in \mathcal{Q} \times (0, \infty)$ of the likelihood function for $\mathcal{E}_j = Y_j - f(t_j, \vec{q})$ which is defined by

$$L(\vec{q}, \sigma^2|\vec{Y}) = \prod_{j=1}^{n} \frac{1}{\sqrt{2\pi\sigma^2}} \exp\left\{-\frac{1}{2\sigma^2}[Y_j - f(t_j, \vec{q})]^2\right\}. \tag{3.45}$$

The resulting solutions q_{MLE} and σ_{MLE}^2 are the maximum likelihood **estimators** (MLEs) for $\vec{q}_0$ and σ_0^2, respectively. We point out that these solutions

$q_{\text{MLE}} = q^n_{\text{MLE}}(\vec{Y})$ and $\sigma^2_{\text{MLE}} = \sigma^{2\,n}_{\text{MLE}}(\vec{Y})$ are *random variables* by virtue of the fact that $\vec{Y}$ is a random variable. The corresponding maximum likelihood **estimates** are obtained by maximizing (3.45) with $\vec{Y} = (Y_1, \ldots, Y_n)^T$ replaced by a given realization $\vec{y} = (y_1, \ldots, y_n)^T$ and will be denoted by $\hat{q}_{\text{MLE}} = \hat{q}^n_{\text{MLE}}$ and $\hat{\sigma}_{\text{MLE}} = \hat{\sigma}^n_{\text{MLE}}$, respectively. In our discussions here and below, almost every quantity of interest is dependent on n, the *size of the set of observations* or the *sampling size*. On occasion, we will express this dependence explicitly by use of superscripts or subscripts, especially when we wish to remind the reader of this dependence. However, for notational convenience we will often suppress the notation of explicit dependence on n.

Maximizing (3.45) is equivalent to maximizing the log likelihood

$$\log L(\vec{q}, \sigma^2 | \vec{Y}) = -\frac{n}{2}\log(2\pi) - \frac{n}{2}\log\sigma^2 - \frac{1}{2\sigma^2}\sum_{j=1}^{n}[Y_j - f(t_j, \vec{q})]^2. \quad (3.46)$$

We determine the maximum of (3.46) by differentiating with respect to $\vec{q}$ (with σ^2 fixed) and with respect to σ^2 (with $\vec{q}$ fixed), setting the resulting equations equal to zero and solving for $\vec{q}$ and σ^2. With σ^2 fixed we solve $\frac{\partial}{\partial \vec{q}}\log L(\vec{q}, \sigma^2 | \vec{Y}) = 0$ which is equivalent to

$$\sum_{j=1}^{n}[Y_j - f(t_j, \vec{q})]\nabla f(t_j, \vec{q}) = 0, \quad (3.47)$$

where as usual $\nabla f = \frac{\partial}{\partial \vec{q}} f = f_{\vec{q}}$. We see that solving (3.47) is the same as the least squares optimization

$$q_{\text{MLE}}(\vec{Y}) = \arg\min_{\vec{q} \in \mathcal{Q}} J(\vec{Y}, \vec{q}) \equiv \arg\min_{\vec{q} \in \mathcal{Q}} \sum_{j=1}^{n}[Y_j - f(t_j, \vec{q})]^2. \quad (3.48)$$

We next fix $\vec{q}$ to be q_{MLE} and solve $\frac{\partial}{\partial \sigma^2}\log L(q_{\text{MLE}}, \sigma^2 | \vec{Y}) = 0$, which yields

$$\sigma^2_{\text{MLE}}(\vec{Y}) = \frac{1}{n}J(\vec{Y}, q_{\text{MLE}}). \quad (3.49)$$

Note that we can solve for q_{MLE} and σ^2_{MLE} separately — a desirable feature, but one that does not arise in more complicated formulations discussed below. The second derivative test (the calculation is deferred to an exercise below) can be used to verify that the expressions above for q_{MLE} and σ^2_{MLE} do indeed maximize (3.46).

However, if we have a vector of observations for the jth covariate t_j, then the statistical model is reformulated as

$$\vec{Y}_j = \vec{f}(t_j, \vec{q}_0) + \vec{\mathcal{E}}_j, \quad (3.50)$$

where $\vec{f} \in R^m$ and

$$V_0 = \text{var}(\vec{\mathcal{E}}_j) = \text{diag}(\sigma^2_{0,1}, \ldots, \sigma^2_{0,m}) \quad (3.51)$$

for $j = 1, \ldots, n$. In this setting, we have allowed for the possibility that the observation coordinates Y_j^i may have different *constant* variances $\sigma_{0,i}^2$, i.e., $\sigma_{0,i}^2$ does not necessarily have to equal $\sigma_{0,k}^2$. If (again) there is sufficient evidence to claim the errors are independent and identically distributed and generated by a normal distribution, then $\vec{\mathcal{E}}_j \sim \mathcal{N}_m(0, V_0)$. We can thus obtain the maximum likelihood estimators $q_{\mathrm{MLE}}(\{\vec{Y}_j\})$ and $V_{\mathrm{MLE}}(\{\vec{Y}_j\})$ for q_0 and V_0 by determining the maximum of the log of the likelihood function for $\vec{\mathcal{E}}_j = \vec{Y}_j - \vec{f}(t_j, \vec{q})$ defined by

$$
\log L(\vec{q}, V | \{Y_j^1, \ldots, Y_j^m\}) = -\frac{n}{2} \sum_{i=1}^{m} \log \sigma_{0,i}^2 - \frac{1}{2} \sum_{i=1}^{m} \frac{1}{\sigma_{0,i}^2} \sum_{j=1}^{n} [Y_j^i - f^i(t_j, \vec{q})]^2
$$

$$
= -\frac{n}{2} \sum_{i=1}^{m} \log \sigma_{0,i}^2 - \sum_{j=1}^{n} [\vec{Y}_j - \vec{f}(t_j, \vec{q})]^T V^{-1} [\vec{Y}_j - \vec{f}(t_j, \vec{q})].
$$

Using arguments similar to those given for the scalar case, we determine the maximum likelihood estimators for $\vec{q}_0$ and V_0 to be

$$
q_{\mathrm{MLE}} = \arg\min_{\vec{q} \in Q} \sum_{j=1}^{n} [\vec{Y}_j - \vec{f}(t_j, \vec{q})]^T V_{\mathrm{MLE}}^{-1} [\vec{Y}_j - \vec{f}(t_j, \vec{q})] \tag{3.52}
$$

$$
V_{\mathrm{MLE}} = \mathrm{diag}\left(\frac{1}{n} \sum_{j=1}^{n} [\vec{Y}_j - \vec{f}(t_j, q_{\mathrm{MLE}})][\vec{Y}_j - \vec{f}(t_j, q_{\mathrm{MLE}})]^T \right). \tag{3.53}
$$

Unfortunately, this is a coupled system, which requires some care when solving numerically. We will discuss this issue further in Sections 3.2.6 and 3.2.9 below.

3.2.4 Unspecified Error Distributions and Asymptotic Theory

In Section 3.2.3 we examined the estimates of $\vec{q}_0$ and V_0 under the assumption *that the error is known and in particular is normally distributed, independent and has constant variance longitudinally.* But what if it is suspected that the error is not normally distributed, or more generally (as in most applications) the error distribution is <u>unknown</u> to the modeler beyond the assumptions on $E[\vec{Y}_j]$ embodied in the model and the assumptions made on $\mathrm{var}(\vec{\mathcal{E}}_j)$? How should we proceed in estimating $\vec{q}_0$ and σ_0 (or V_0) in these circumstances? In the next several sections we will review two estimation procedures for such situations: ordinary least squares (OLS) and generalized least squares (GLS).

3.2.5   Ordinary Least Squares (OLS)

The statistical model in the scalar case takes the form

$$Y_j = f(t_j, \vec{q}_0) + \mathcal{E}_j, \tag{3.54}$$

where the variance $\text{var}(\mathcal{E}_j) = \sigma_0^2$ is assumed constant in longitudinal data (note that the error's distribution is not specified). We also note that the assumption that the observation errors are uncorrelated across j (i.e., time) may be a reasonable one when the observations are taken with sufficient intermittency or when the primary source of error is measurement error. If we define

$$q_{\text{OLS}}(\vec{Y}) = q_{\text{OLS}}^n(\vec{Y}) = \arg\min_{\vec{q} \in \mathcal{Q}} \sum_{j=1}^{n} [Y_j - f(t_j, \vec{q})]^2, \tag{3.55}$$

then q_{OLS} can be viewed as minimizing the distance between the data and model where all observations are treated as of equal importance. We note that minimizing the functional in (3.55) corresponds to solving for $\vec{q}$ in

$$\sum_{j=1}^{n} [Y_j - f(t_j, \vec{q})] \nabla f(t_j, \vec{q}) = 0. \tag{3.56}$$

We point out that q_{OLS} is a *random variable* (because $\mathcal{E}_j = Y_j - f(t_j, \vec{q})$ is a random variable); hence if $\{y_j\}_{j=1}^n$ is a realization of the *random process* $\{Y_j\}_{j=1}^n$ then solving

$$\hat{q}_{\text{OLS}} = \hat{q}_{\text{OLS}}^n = \arg\min_{\vec{q} \in \mathcal{Q}} \sum_{j=1}^{n} [y_j - f(t_j, \vec{q})]^2 \tag{3.57}$$

provides a realization for q_{OLS}. (A remark on notation: for a random variable or estimator q, we will always denote a corresponding realization or estimate with an <u>over hat</u>, e.g., $\hat{q}$ is an estimate for q so that we have here abandoned the usual convention of capital letters for random variables and a lower case letter for a corresponding realization — this again follows convention in the statistical literature.)

Noting that

$$\sigma_0^2 = \frac{1}{n} E\left[\sum_{j=1}^{n} [Y_j - f(t_j, \vec{q}_0)]^2 \right] \tag{3.58}$$

suggests that once we have solved for q_{OLS} in (3.55), we may readily obtain an estimate $\hat{\sigma}_{\text{OLS}}^2$ ($= \hat{\sigma}_{\text{MLE}}^2{}^n$ – see (3.49)) for σ_0^2.

Even though the error's distribution is not specified, we can use asymptotic theory to approximate the mean and variance of the random variable q_{OLS} [17]. As will be explained in more detail below, as $n \to \infty$, we have that

$$q_{\text{OLS}} = q_{\text{OLS}}^n \sim \mathcal{N}_p(\vec{q}_0, \Sigma_0^n) \approx \mathcal{N}_p(\vec{q}_0, \sigma_0^2 [\chi^{nT}(\vec{q}_0) \chi^n(\vec{q}_0)]^{-1}), \tag{3.59}$$

where the sensitivity matrix $\chi(\vec{q}) = \chi^n(\vec{q}) = \{\chi^n_{jk}\}$ is defined as

$$\chi^n_{jk}(\vec{q}) = \frac{\partial f(t_j, \vec{q})}{\partial q_k}, \quad j = 1, \ldots, n, \quad k = 1, \ldots, p,$$

and

$$\Sigma^n_0 \equiv \sigma^2_0 [n\Omega_0]^{-1} \tag{3.60}$$

with

$$\Omega_0 \equiv \lim_{n \to \infty} \frac{1}{n} \chi^{nT}(\vec{q}_0) \chi^n(\vec{q}_0), \tag{3.61}$$

where the limit is assumed to exist (see [17]). However, $\vec{q}_0$ and σ^2_0 are generally unknown, so one usually will use instead the *realization* $\vec{y} = (y_1, \ldots, y_n)^T$ of the random process $\vec{Y}$ to obtain the estimate

$$\hat{q}_{\text{OLS}} = \arg\min_{\vec{q} \in \mathcal{Q}} \sum_{j=1}^{n} [y_j - f(t_j, \vec{q})]^2 \tag{3.62}$$

and the *bias adjusted* estimate

$$\hat{\sigma}^2_{\text{OLS}} = \frac{1}{n - p} \sum_{j=1}^{n} [y_j - f(t_j, \hat{q})]^2 \tag{3.63}$$

to use as an approximation in (3.59).

We note that (3.63) represents the estimate for σ^2_0 of (3.58) with the factor $\frac{1}{n}$ replaced by the factor $\frac{1}{n-p}$ (in the linear case the estimate with $\frac{1}{n}$ can be shown to be biased downward and the same behavior can be observed in the general nonlinear case (see Chap. 12 of [17] and p. 28 of [11])). We remark that (3.58) is true even in the general nonlinear case (it does not rely on any asymptotic theories although it does depend on the assumption of constant variance being correct).

Both $\hat{q} = \hat{q}_{\text{OLS}}$ and $\hat{\sigma}^2 = \hat{\sigma}^2_{\text{OLS}}$ will then be used to approximate the covariance matrix

$$\Sigma^n_0 \approx \hat{\Sigma}^n \equiv \hat{\sigma}^2 [\chi^{nT}(\hat{q}) \chi^n(\hat{q})]^{-1}. \tag{3.64}$$

We can obtain the standard errors $SE(\hat{q}_{\text{OLS},k})$ (discussed in more detail in the next section) for the kth element of $\hat{q}_{\text{OLS}}$ by calculating $SE(\hat{q}_{\text{OLS},k}) \approx \sqrt{\hat{\Sigma}^n_{kk}}$. Also note the similarity between the MLE equations (3.48) and (3.49), and the scalar OLS equations (3.62) and (3.63). That is, under a normality assumption for the error, the MLE and OLS formulations are equivalent.

However, if we have a vector of observations for the jth covariate t_j and we assume the variance is still constant in longitudinal data, then the statistical model is reformulated as

$$\vec{Y}_j = \vec{f}(t_j, \vec{q}_0) + \vec{\mathcal{E}}_j, \tag{3.65}$$

where $\vec{f} \in R^m$ and

$$V_0 = \text{var}(\vec{\mathcal{E}}_j) = \text{diag}(\sigma_{0,1}^2, \ldots, \sigma_{0,m}^2) \qquad (3.66)$$

for $j = 1, \ldots, n$. Just as in the MLE case, we have allowed for the possibility that the observation coordinates Y_j^i may have different *constant* variances $\sigma_{0,i}^2$, i.e., $\sigma_{0,i}^2$ does not necessarily have to equal $\sigma_{0,k}^2$. We note that this formulation also can be used to treat the case where V_0 is used to simply scale the observations, i.e., $V_0 = \text{diag}(v_1, \ldots, v_m)$ is known. In this case the formulation is simply a *vector OLS* (sometimes also called a *weighted least squares* (WLS)). The problem will consist of finding the minimizer

$$q_{\text{OLS}} = \arg\min_{\vec{q} \in \mathcal{Q}} \sum_{j=1}^{n} [\vec{Y}_j - \vec{f}(t_j, \vec{q})]^T V_0^{-1} [\vec{Y}_j - \vec{f}(t_j, \vec{q})], \qquad (3.67)$$

where the procedure weights elements of the vector $\vec{Y}_j - \vec{f}(t_j, \vec{q})$ according to their variability. (Some authors refer to (3.67) as a generalized least squares (GLS) procedure, but we will make use of this terminology in a different formulation in subsequent discussions.) Just as in the scalar OLS case, q_{OLS} is a *random variable* (again because $\vec{\mathcal{E}}_j = \vec{Y}_j - \vec{f}(t_j, \vec{q})$ is); hence if $\{\vec{y}_j\}_{j=1}^n$ is a realization of the *random process* $\{\vec{Y}_j\}_{j=1}^n$ then solving

$$\hat{q}_{\text{OLS}} = \arg\min_{\vec{q} \in \mathcal{Q}} \sum_{j=1}^{n} [\vec{y}_j - \vec{f}(t_j, \vec{q})]^T V_0^{-1} [\vec{y}_j - \vec{f}(t_j, \vec{q})] \qquad (3.68)$$

provides an estimate (realization) $\hat{q} = \hat{q}_{\text{OLS}}$ for q_{OLS}. By the definition of variance

$$V_0 = \text{diag } E \left(\frac{1}{n} \sum_{j=1}^{n} [\vec{Y}_j - \vec{f}(t_j, \vec{q}_0)][\vec{Y}_j - \vec{f}(t_j, \vec{q}_0)]^T \right),$$

so an unbiased estimate of V_0 for the realization $\{\vec{y}_j\}_{j=1}^n$ is

$$\hat{V} = \text{diag} \left(\frac{1}{n-p} \sum_{j=1}^{n} [\vec{y}_j - \vec{f}(t_j, \hat{q})][\vec{y}_j - \vec{f}(t_j, \hat{q})]^T \right). \qquad (3.69)$$

However, the estimate $\hat{q}$ requires the (generally unknown) matrix V_0, and V_0 requires the unknown vector $\vec{q}_0$, so we will instead use the following expressions to calculate $\hat{q}$ and $\hat{V}$:

$$\vec{q}_0 \approx \hat{q} = \arg\min_{\vec{q} \in \mathcal{Q}} \sum_{j=1}^{n} [\vec{y}_j - \vec{f}(t_j, \vec{q})]^T \hat{V}^{-1} [\vec{y}_j - \vec{f}(t_j, \vec{q})] \qquad (3.70)$$

$$V_0 \approx \hat{V} = \text{diag} \left(\frac{1}{n-p} \sum_{j=1}^{n} [\vec{y}_j - \vec{f}(t_j, \hat{q})][\vec{y}_j - \vec{f}(t_j, \hat{q})]^T \right). \qquad (3.71)$$

Note that the expressions for $\hat{q}$ and $\hat{V}$ constitute a coupled system of equations that will require greater effort in implementing a numerical scheme.

Just as in the scalar case, we can determine the asymptotic properties of the OLS estimator (3.67). As $n \to \infty$, q_{OLS} has the following asymptotic properties [11, 17]:

$$q_{\text{OLS}} \sim \mathcal{N}_p(\vec{q}_0, \Sigma_0^n), \tag{3.72}$$

where

$$\Sigma_0^n \approx \left(\sum_{j=1}^n D_j^T(\vec{q}_0) V_0^{-1} D_j(\vec{q}_0) \right)^{-1}, \tag{3.73}$$

and the $m \times p$ matrix $D_j(\vec{q}) = D_j^n(\vec{q})$ is given by

$$\begin{pmatrix} \frac{\partial f_1(t_j, \vec{q})}{\partial q_1} & \frac{\partial f_1(t_j, \vec{q})}{\partial q_2} & \cdots & \frac{\partial f_1(t_j, \vec{q})}{\partial q_p} \\ \vdots & \vdots & & \vdots \\ \frac{\partial f_m(t_j, \vec{q})}{\partial q_1} & \frac{\partial f_m(t_j, \vec{q})}{\partial q_2} & \cdots & \frac{\partial f_m(t_j, \vec{q})}{\partial q_p} \end{pmatrix}.$$

Since the true value of the parameters $\vec{q}_0$ and V_0 are unknown, their estimates $\hat{q}$ and $\hat{V}$ are used to approximate the asymptotic properties of the least squares estimator q_{OLS}:

$$q_{\text{OLS}} \sim \mathcal{N}_p(\vec{q}_0, \Sigma_0^n) \approx \mathcal{N}_p(\hat{q}, \hat{\Sigma}^n), \tag{3.74}$$

where

$$\Sigma_0^n \approx \hat{\Sigma}^n = \left(\sum_{j=1}^n D_j^T(\hat{q}) \hat{V}^{-1} D_j(\hat{q}) \right)^{-1}. \tag{3.75}$$

The standard errors $SE(\hat{q}_{\text{OLS},k})$ can then be calculated for the kth element of $\hat{q}_{\text{OLS}}$ by $SE(\hat{q}_{\text{OLS},k}) \approx \sqrt{\hat{\Sigma}_{kk}}$. Again, we point out the similarity between the MLE equations (3.52) and (3.53), and the OLS equations (3.70) and (3.71) for the vector statistical model (3.65).

3.2.6 Numerical Implementation of the Vector OLS Procedure

In the scalar statistical model (3.54), the estimates $\hat{q}$ and $\hat{\sigma}$ can be solved for separately (this is also true of the vector OLS in the case $V_0 = \sigma_0^2 I_m$, where I_m is the $m \times m$ identity matrix) and thus the numerical implementation is straightforward — first determine $\hat{q}_{\text{OLS}}$ according to (3.62) and then calculate

$\hat{\sigma}^2_{\text{OLS}}$ according to (3.63). However, the estimates $\hat{q}$ and $\hat{V}$ in the case of the vector statistical model (3.65) require more effort since they are coupled:

$$\hat{q} = \arg\min_{\vec{q} \in \mathcal{Q}} \sum_{j=1}^{n} [\vec{y}_j - \vec{f}(t_j, \vec{q})]^T \hat{V}^{-1} [\vec{y}_j - \vec{f}(t_j, \vec{q})] \tag{3.76}$$

$$\hat{V} = \text{diag}\left(\frac{1}{n-p} \sum_{j=1}^{n} [\vec{y}_j - \vec{f}(t_j, \hat{q})][\vec{y}_j - \vec{f}(t_j, \hat{q})]^T \right). \tag{3.77}$$

To solve this coupled system the following iterative process will be followed:

1. Set $\hat{V}^{(0)} = \mathbf{I}$ and solve for the initial estimate $\hat{q}^{(0)}$ using (3.76). Set $k = 0$.

2. Use $\hat{q}^{(k)}$ to calculate $\hat{V}^{(k+1)}$ using (3.77).

3. Re-estimate $\vec{q}$ by solving (3.76) with $\hat{V} = \hat{V}^{(k+1)}$ to obtain $\hat{q}^{(k+1)}$.

4. Set $k = k + 1$ and return to step 2. Terminate the process and set $\hat{q}_{\text{OLS}} = \hat{q}^{(k+1)}$ when two successive estimates for $\hat{q}$ are sufficiently close to one another.

3.2.7   Generalized Least Squares (GLS)

Although in Section 3.2.5 the error's distribution remained unspecified, we did however require that the error remain constant in variance in longitudinal data. That assumption may not be appropriate for data sets whose error is not constant in a longitudinal sense. A common *relative error* model (e.g., one in which the size of the observation error is assumed proportional to the size of the observed quantity, an assumption which might be reasonable when counting individuals in a population) that experimentalists use in this instance for the scalar observation case [11] is

$$Y_j = f(t_j, \vec{q}_0)(1 + \mathcal{E}_j), \tag{3.78}$$

where $E(Y_j) = f(t_j, \vec{q}_0)$ and $\text{var}(Y_j) = \sigma_0^2 f^2(t_j, \vec{q}_0)$ which derives from the assumptions that $E[\mathcal{E}_j] = 0$ and $\text{var}(\mathcal{E}_j) = \sigma_0^2$. We see that the variance generated in this fashion is model dependent and hence generally is longitudinally non-constant variance. The method we will use to estimate $\vec{q}_0$ and σ_0^2 can be viewed as a particular form of the Generalized Least Squares (GLS) method.

To define the *random variable* q_{GLS}, the following equation must be solved for the estimator q_{GLS}:

$$\sum_{j=1}^{n} w_j [Y_j - f(t_j, q_{\text{GLS}})] \nabla f(t_j, q_{\text{GLS}}) = 0, \tag{3.79}$$

where Y_j obeys (3.78) and $w_j = f^{-2}(t_j, q_{\mathrm{GLS}})$. We note these are the so-called normal equations (obtained by equating the gradient of the weighted least squares criterion to zero in the case the weights w_j are independent of q). The quantity q_{GLS} is a random variable, hence if $\{y_j\}_{j=1}^n$ is a *realization* of the random process Y_j, then solving

$$\sum_{j=1}^n f^{-2}(t_j, \hat{q})[y_j - f(t_j, \hat{q})]\nabla f(t_j, \hat{q}) = 0 \qquad (3.80)$$

for $\hat{q}$ we obtain an estimate $\hat{q}_{\mathrm{GLS}}$ for q_{GLS}.

The GLS estimator $q_{\mathrm{GLS}} = q_{\mathrm{GLS}}^n$ has the following asymptotic properties [11]:

$$q_{\mathrm{GLS}} \sim \mathcal{N}_p(\vec{q}_0, \Sigma_0^n), \qquad (3.81)$$

where

$$\Sigma_0^n \approx \sigma_0^2 \left(F_{\vec{q}}^T(\vec{q}_0)W(\vec{q}_0)F_{\vec{q}}(\vec{q}_0)\right)^{-1}, \qquad (3.82)$$

$$F_{\vec{q}}(\vec{q}) = F_{\vec{q}}^n(\vec{q}) = \begin{pmatrix} \frac{\partial f(t_1,\vec{q})}{\partial q_1} & \frac{\partial f(t_1,\vec{q})}{\partial q_2} & \cdots & \frac{\partial f(t_1,\vec{q})}{\partial q_p} \\ \vdots & & & \vdots \\ \frac{\partial f(t_n,\vec{q})}{\partial q_1} & \frac{\partial f(t_n,\vec{q})}{\partial q_2} & \cdots & \frac{\partial f(t_n,\vec{q})}{\partial q_p} \end{pmatrix} = \begin{pmatrix} \nabla f(t_1,\vec{q})^T \\ \vdots \\ \nabla f(t_n,\vec{q})^T \end{pmatrix},$$

and $W^{-1}(\vec{q}) = \operatorname{diag}\left(f^2(t_1, \vec{q}), \ldots, f^2(t_n, \vec{q})\right)$. Note that because $\vec{q}_0$ and σ_0^2 are unknown, the estimates $\hat{q} = \hat{q}_{\mathrm{GLS}}$ and $\hat{\sigma}^2 = \hat{\sigma}_{\mathrm{GLS}}^2$ will be used in (3.82) to calculate

$$\Sigma_0^n \approx \hat{\Sigma}^n = \hat{\sigma}^2 \left(F_{\vec{q}}^T(\hat{q})W(\hat{q})F_{\vec{q}}(\hat{q})\right)^{-1},$$

where [11] we take the approximation

$$\sigma_0^2 \approx \hat{\sigma}_{\mathrm{GLS}}^2 = \frac{1}{n-p}\sum_{j=1}^n \frac{1}{f^2(t_j, \hat{q})}[y_j - f(t_j, \hat{q})]^2.$$

We can then approximate the standard errors of $\hat{q}_{\mathrm{GLS}}$ by taking the square roots of the diagonal elements of $\hat{\Sigma}$. We will also mention that the solutions to (3.70) and (3.80) depend upon the numerical method used to find the minimum or root, and since Σ_0 depends upon the estimate for $\vec{q}_0$, the standard errors are therefore affected by the numerical method chosen.

3.2.8 GLS Motivation

We note the similarity between (3.56) and (3.80). The GLS equation (3.80) can be motivated by examining the weighted least squares (WLS) estimator

$$q_{\mathrm{WLS}} = \arg\min_{\vec{q}\in Q} \sum_{j=1}^n w_j[Y_j - f(t_j, \vec{q})]^2. \qquad (3.83)$$

In many situations where the observation process is well understood, the weights $\{w_j\}$ may be known. The WLS estimate can be thought of minimizing the distance between the data and model while taking into account unequal quality of the observations [11]. If we differentiate the sum of squares in (3.83) with respect to $\vec{q}$ and *then* choose $w_j = f^{-2}(t_j, \vec{q})$, an estimate $\hat{q}_{\text{GLS}}$ is obtained by solving

$$\sum_{j=1}^{n} w_j [y_j - f(t_j, \vec{q})] \nabla f(t_j, \vec{q}) = 0$$

for $\vec{q}$. However, we note the GLS relationship (3.80) does _not_ follow from minimizing the weighted least squares with weights chosen as $w_j = f^{-2}(t_j, \vec{q})$.

Another motivation for the GLS estimating equation (3.80) can be found in [9]. In the text, the authors claim that if the errors (and hence the data) are distributed according to the gamma distribution, then the maximum likelihood estimator for $\vec{q}$ is the solution to

$$\sum_{j=1}^{n} f^{-2}(t_j, \vec{q})[Y_j - f(t_j, \vec{q})] \nabla f(t_j, \vec{q}) = 0,$$

which is equivalent to (3.80). The connection between the MLE and our GLS method is reassuring, but it also poses another interesting question: what if the variance of the data is assumed to be independent of the model output $f(t_j, \vec{q})$ but depends on some other function $g(t_j, \vec{q})$ (i.e., $\text{var}(Y_j) = \sigma_0^2 g^2(t_j, \vec{q}) = \sigma_0^2/w_j$)? Is there a corresponding maximum likelihood estimator of $\vec{q}$ whose form is equivalent to the appropriate GLS estimating equation $(w_j = g^{-2}(t_j, \vec{q}))$

$$\sum_{j=1}^{n} g^{-2}(t_j, \vec{q})[Y_j - f(t_j, \vec{q})] \nabla f(t_j, \vec{q}) = 0 \ ? \tag{3.84}$$

In their text, Carroll and Ruppert [9] briefly describe how distributions belonging to the exponential family of distributions generate maximum-likelihood estimating equations equivalent to (3.84).

3.2.9 Numerical Implementation of the GLS Procedure

Recall that an estimate $\hat{q}_{\text{GLS}}$ can either be solved for directly according to (3.80) or iteratively using the equations outlined in Section 3.2.7. The iterative procedure as described in [11] is summarized below:

1. Estimate $\hat{q}_{\text{GLS}}$ by $\hat{q}^{(0)}$ using the OLS equation (3.55). Set $k = 0$.

2. Form the weights $\hat{w}_j = f^{-2}(t_j, \hat{q}^{(k)})$.

3. Re-estimate $\hat{q}$ by solving

$$\hat{q}^{(k+1)} = \arg\min_{q \in \mathcal{Q}} \sum_{j=1}^{n} \hat{w}_j \left(y_j - f(t_j, \vec{q})\right)^2$$

to obtain the $k+1$ estimate $\hat{q}^{(k+1)}$ for $\hat{q}_{\text{GLS}}$.

4. Set $k = k + 1$ and return to step 2. Terminate the process when two of the successive estimates for $\hat{q}_{\text{GLS}}$ are sufficiently close.

We note that the above iterative procedure was formulated by minimizing (over $\vec{q} \in \mathcal{Q}$)

$$\sum_{j=1}^{n} f^{-2}(t_j, \tilde{q})[y_j - f(t_j, \vec{q})]^2$$

and then updating the weights $w_j = f^{-2}(t_j, \tilde{q})$ after each iteration. One would hope that after a sufficient number of iterations $\hat{w}_j$ would converge to $f^{-2}(t_j, \hat{q}_{\text{GLS}})$. Fortunately, under reasonable conditions, if the process enumerated above is continued a sufficient number of times [11], then $\hat{w}_j \to f^{-2}(t_j, \hat{q}_{\text{GLS}})$.

3.3   Computation of $\hat{\Sigma}^n$, Standard Errors and Confidence Intervals

We return to the case of n scalar longitudinal observations and consider the OLS case of Section 3.2.5 (the extension of these ideas to vectors is completely straight-forward). These n scalar observations are represented by the statistical model

$$Y_j \equiv f(t_j, \vec{q}_0) + \mathcal{E}_j, \quad j = 1, 2, \ldots, n, \tag{3.85}$$

where $f(t_j, \vec{q}_0)$ is the model for the observations in terms of the state variables and $\vec{q}_0 \in R^p$ is a set of theoretical "true" parameter values (assumed to exist in a standard statistical approach). We further assume that the errors $\mathcal{E}_j$, $j = 1, 2, \ldots, n$, are independent identically distributed (*i.i.d.*) random variables with mean $E[\mathcal{E}_j] = 0$ and constant variance $\text{var}(\mathcal{E}_j) = \sigma_0^2$, where σ_0^2 is unknown. The observations Y_j are then *i.i.d.* with mean $E[Y_j] = f(t_j, \vec{q}_0)$ and variance $\text{var}(Y_j) = \sigma_0^2$.

Recall that in the ordinary least squares (OLS) approach, we seek to use a realization $\{y_j\}$ of the observation process $\{Y_j\}$ along with the model to determine a vector $\hat{q}_{\text{OLS}}^n$ where

$$\hat{q}_{\text{OLS}}^n = \arg\min J_n(\vec{q}) = \sum_{j=1}^{n} [y_j - f(t_j, \vec{q})]^2. \tag{3.86}$$

Since Y_j is a random variable, the corresponding estimator $q^n = q^n_{OLS}$ (here we wish to emphasize the dependence on the sample size n) is also a random variable with a distribution called the *sampling distribution*. Knowledge of this sampling distribution provides uncertainty information (e.g., standard errors) for the numerical values of $\hat{q}^n$ obtained using a specific data set $\{y_j\}$. In particular, loosely speaking, the sampling distribution characterizes the distribution of possible values the estimator could take on across all possible realizations with data of size n that could be collected. The standard errors thus approximate the extent of variability in possible values across all possible realizations, and hence provide a measure of the extent of uncertainty involved in estimating q using the specific estimator and sample size n in actual data collection.

Under reasonable assumptions on smoothness and regularity (the smoothness requirements for model solutions are readily verified using continuous dependence results for differential equations in most examples; the regularity requirements include, among others, conditions on *how the observations are taken* as sample size increases, i.e., as $n \to \infty$), the standard nonlinear regression approximation theory ([11, 13, 15], and Chapter 12 of [17]) for **asymptotic (as $n \to \infty$) distributions** can be invoked. As stated above, this theory yields that the sampling distributions for the estimators $q^n(\vec{Y})$, where $\vec{Y} = (Y_1, \ldots, Y_n)^T$, can be approximated by a p-multivariate Gaussian (i.e., the sequence of cumulative distribution functions converge as $n \to \infty$ at points of continuity of the limit cdf — this is called *convergence in distribution*) with mean $E[q^n(\vec{Y})] \approx \vec{q}_0$ and covariance matrix $\text{var}(q^n(\vec{Y})) \approx \Sigma_0^n = \sigma_0^2 [n\Omega_0]^{-1} \approx \sigma_0^2 [\chi^{nT}(\vec{q}_0)\chi^n(\vec{q}_0)]^{-1}$. Here $\chi^n(\vec{q}) = F_{\vec{q}}(\vec{q})$ is the $n \times p$ sensitivity matrix with elements

$$\chi_{jk}(\vec{q}) = \frac{\partial f(t_j, \vec{q})}{\partial q_k} \qquad \text{and} \qquad F_{\vec{q}}(\vec{q}) \equiv (f_{1\vec{q}}(\vec{q}), \ldots, f_{n\vec{q}}(\vec{q}))^T,$$

where $f_{j\vec{q}}(\vec{q}) = \frac{\partial f}{\partial \vec{q}}(t_j, \vec{q})$. That is, for n large, the sampling distribution approximately satisfies

$$q^n_{OLS}(\vec{Y}) \sim \mathcal{N}_p(\vec{q}_0, \Sigma_0^n) \approx \mathcal{N}_p(\vec{q}_0, \sigma_0^2[\chi^{nT}(\vec{q}_0)\chi^n(\vec{q}_0)]^{-1}). \qquad (3.87)$$

There are typically several ways to compute the matrix $F_{\vec{q}}$ (which are actually the well known sensitivity functions widely used in applied mathematics and engineering (see the discussions in [2] and the references therein)). First, the elements of the matrix $\chi = (\chi_{jk})$ can always be estimated using the forward difference

$$\chi_{jk}(\vec{q}) = \frac{\partial f(t_j, \vec{q})}{\partial q_k} \approx \frac{f(t_j, \vec{q} + h_k) - f(t_j, \vec{q})}{|h_k|},$$

where h_k is a p-vector with a nonzero entry in only the kth component. But, of course, the choice of h_k can be problematic in practice.

Alternatively, if the $f(t_j, \vec{q})$ correspond to longitudinal observations $\vec{y}(t_j) = C\vec{z}(t_j; \vec{q})$ of solutions $\vec{z} \in R^N$ to a parameterized N-vector differential equation system $\dot{\vec{z}} = \vec{g}(t, \vec{z}(t), \vec{q})$ as in (3.38), then one can use the $N \times p$ matrix **sensitivity equations** (see [3, 4] and the references therein)

$$\frac{d}{dt}\left(\frac{\partial \vec{z}}{\partial \vec{q}}\right) = \frac{\partial \vec{g}}{\partial \vec{z}}\frac{\partial \vec{z}}{\partial \vec{q}} + \frac{\partial \vec{g}}{\partial \vec{q}} \tag{3.88}$$

to obtain

$$\frac{\partial f(t_j, \vec{q})}{\partial q_k} = C\frac{\partial \vec{z}(t_j, \vec{q})}{\partial q_k}.$$

Finally, in some cases the function $f(t_j, \vec{q})$ may be sufficiently simple so as to allow one to derive analytical expressions for the components of $F_{\vec{q}}$.

We remark that often one also wants to include initial conditions as part of the unknown vector $\vec{q}$ to be estimated. In this case, one can readily derive sensitivity equations for sensitivities with respect to initial conditions that are analogous to (3.88). See [2] for examples.

Since $\vec{q}_0, \sigma_0$ are unknown, we will use their estimates to make the approximation

$$\Sigma_0^n \approx \sigma_0^2[\chi^{nT}(\vec{q}_0)\chi^n(\vec{q}_0)]^{-1} \approx \hat{\Sigma}^n(\hat{q}_{\text{OLS}}^n) = \hat{\sigma}^2[\chi^{nT}(\hat{q}_{\text{OLS}}^n)\chi^n(\hat{q}_{\text{OLS}}^n)]^{-1}, \tag{3.89}$$

where the approximation $\hat{\sigma}^2$ to σ_0^2, as discussed earlier, is given by

$$\sigma_0^2 \approx \hat{\sigma}^2 = \frac{1}{n-p}\sum_{j=1}^{n}[y_j - f(t_j, \hat{q}_{\text{OLS}}^n)]^2. \tag{3.90}$$

Standard errors to be used in the confidence interval calculations are thus given by $SE_k(\hat{q}^n) = \sqrt{\Sigma_{kk}(\hat{q}^n)}$, $k = 1, 2, \ldots, p$ (see [10]).

In order to compute the confidence intervals (at the $100(1 - \alpha)\%$ level) for the estimated parameters in our example, we define the confidence level parameters associated with the estimated parameters so that

$$P\{\hat{q}_k^n - t_{1-\alpha/2}SE_k(\hat{q}^n) < q_{0k} < \hat{q}_k^n + t_{1-\alpha/2}SE_k(\hat{q}^n)\} = 1 - \alpha, \tag{3.91}$$

where $\alpha \in [0, 1]$ and $t_{1-\alpha/2} \in R_+$. Given a small α value (e.g., $\alpha = .05$ for 95% confidence intervals), the critical value $t_{1-\alpha/2}$ is computed from the Student's t distribution t^{n-p} with $n - p$ degrees of freedom. The value of $t_{1-\alpha/2}$ is determined by $P\{T \geq t_{1-\alpha/2}\} = \alpha/2$ where $T \sim t^{n-p}$. In general, a confidence interval is constructed so that, if the confidence interval could be constructed for each possible realization of data of size n that could have been collected, $100(1 - \alpha)\%$ of the intervals so constructed would contain the true value q_{0k}. Thus, a confidence interval provides further information on the extent of uncertainty involved in estimating q_0 using the given estimator and sample size n.

When one is taking longitudinal samples corresponding to solutions of a dynamical system, the $n \times p$ sensitivity matrix depends explicitly on where in time the observations are taken when $f(t_j, \vec{q}) = Cz(t_j, \vec{q})$ as mentioned above. That is, the sensitivity matrix

$$\chi(\vec{q}) = F_{\vec{q}}(\vec{q}) = \left(\frac{\partial f(t_j, \vec{q})}{\partial \vec{q}} \right)$$

depends on the number n and the nature (for example, how taken) of the sampling times $\{t_j\}$. Moreover, it is the matrix $[\chi^T \chi]^{-1}$ in (3.89) and the parameter $\hat{\sigma}^2$ in (3.90) that ultimately determine the standard errors and confidence intervals. At first investigation of (3.90), it appears that an increased number n of samples might drive $\hat{\sigma}^2$ (and hence the SE) to zero as long as this is done in a way to maintain a bound on the residual sum of squares in (3.90). However, we observe that the *condition number* of the matrix $\chi^T \chi$ is also very important in these considerations and increasing the sampling could potentially adversely affect the inversion of $\chi^T \chi$. In this regard, we note that among the important hypotheses in the asymptotic statistical theory (see pp. 571 of [17]) is the existence of a matrix function $\Omega(\vec{q})$ such that

$$\frac{1}{n} \chi^{nT}(\vec{q}) \chi^n(\vec{q}) \to \Omega(\vec{q}) \quad \text{uniformly in } \vec{q} \text{ as } n \to \infty,$$

with $\Omega_0 = \Omega(\vec{q}_0)$ being a **nonsingular** matrix. It is this condition that is rather easily violated in practice when one is dealing with data from differential equation systems, especially near an equilibrium or steady state (see the examples of [4]).

All of the above theory readily generalizes to vector systems with partial, non-scalar observations. Suppose now we have the vector system (3.38) with partial vector observations given by equation (3.40). That is, suppose we have m coordinate observations where $m \leq N$. In this case, we have

$$\frac{d\vec{z}}{dt}(t) = \vec{g}(t, \vec{z}(t), \vec{q}) \tag{3.92}$$

and

$$\vec{y}_j = \vec{f}(t_j, \vec{q}_0) + \vec{\epsilon}_j = C\vec{z}(t_j, \vec{q}_0) + \vec{\epsilon}_j, \tag{3.93}$$

where C is an $m \times N$ matrix and $\vec{f} \in R^m, \vec{z} \in R^N$. As already explained in Section 3.2.5, if we assume that different observation coordinates f_i may have different variances σ_i^2 associated with different coordinates of the errors $\mathcal{E}_j$, then we have that $\vec{\mathcal{E}}_j$ is an m-dimensional random vector with

$$E[\vec{\mathcal{E}}_j] = 0, \quad \text{var}(\vec{\mathcal{E}}_j) = V_0,$$

where $V_0 = \text{diag}(\sigma_{0,1}^2, ..., \sigma_{0,m}^2)$, and we may follow a similar asymptotic theory to calculate approximate covariances, standard errors and confidence intervals for parameter estimates.

Since the computations for standard errors and confidence intervals (and also *model comparison tests*) depend on *an asymptotic limit distribution theory*, one should interpret the findings as sometimes crude indicators of uncertainty inherent in the inverse problem findings. Nonetheless, it is useful to consider the formal mathematical requirements underpinning these techniques. We offer the following summary of possibilities:

(1) Among the more readily checked hypotheses are those of the statistical model requiring that the errors $\mathcal{E}_j$, $j = 1, 2, \ldots, n$, are independent and identically distributed (*i.i.d.*) random variables with mean $E[\mathcal{E}_j] = 0$ and constant variance $\mathrm{var}(\mathcal{E}_j) = \sigma_0^2$. After carrying out the estimation procedures, one can readily plot the *residuals* $r_j = y_j - f(t_j, \hat{q}_{OLS}^n)$ *vs. time* t_j and the *residuals vs. the resulting estimated model/observation* $f(t_j, \hat{q}_{OLS}^n)$ *values.* A random pattern for the first is strong support for validity of the independence assumption; a non-increasing, random pattern for the latter suggests the assumption of the constant variance may be reasonable.

(2) The underlying assumption that sampling size n must be large (recall the theory is asymptotic in that it holds as $n \to \infty$) is not so readily "verified" and is often ignored (albeit at the user's peril in regard to the quality of the uncertainty findings). Often asymptotic results provide remarkably good approximations to the true sampling distributions for finite n. However, in practice there is no way to ascertain whether theory holds for a specific example.

3.4 Investigation of Statistical Assumptions

The form of error in the data (which of course is rarely known) dictates which method from those discussed above one should choose. The OLS method is most appropriate for constant variance observations of the form $Y_j = f(t_j, \vec{q}_0) + \mathcal{E}_j$ whereas the GLS should be used for problems in which we have nonconstant variance observations $Y_j = f(t_j, \vec{q}_0)(1 + \mathcal{E}_j)$.

We emphasize that to obtain *the correct standard errors* in an inverse problem calculation, the OLS method (and *corresponding asymptotic formulas*) must be used with constant variance generated data, while the GLS method (and *corresponding asymptotic formulas*) should be applied to nonconstant variance generated data.

Not doing so can lead to *incorrect conclusions*. In either case, the standard error calculations are not valid unless the correct formulas (which depend on the error structure) are employed. Unfortunately, it is very difficult to ascertain the structure of the error, and hence the correct method to use, without *a priori* information. Although the error structure cannot definitively be determined, the two residuals tests can be performed *after* the estimation procedure has been completed to assist in concluding whether or not the correct asymptotic statistics were used.

3.4.1   Residual Plots

One can carry out simulation studies with a proposed mathematical model to assist in understanding the behavior of the model in inverse problems with different types of data with respect to mis-specification of the statistical model. For example, we consider a statistical model with constant variance (CV) noise

$$Y_j = f(t_j, \vec{q}_0) + \frac{\eta}{100} \mathcal{E}_j, \qquad \text{var}(Y_j) = \frac{\eta^2}{10000} \sigma^2,$$

and another with nonconstant variance (NCV) noise

$$Y_j = f(t_j, \vec{q}_0)(1 + \frac{\eta}{100} \mathcal{E}_j), \qquad \text{var}(Y_j) = \frac{\eta^2}{10000} \sigma^2 f^2(t_j, \vec{q}_0).$$

We obtain a data set by considering a *realization* $\{y_j\}_{j=1}^n$ of the random process $\{Y_j\}_{j=1}^n$ through a realization of $\{\mathcal{E}_j\}_{j=1}^n$, and then calculate an estimate $\hat{q}$ of $\vec{q}_0$ using the OLS or GLS procedure.

We will then use the *residuals* $r_j = y_j - f(t_j, \hat{q})$ to test whether the data set is *i.i.d.* and possesses the assumed variance structure. If a data set has constant variance error then

$$Y_j = f(t_j, \vec{q}_0) + \mathcal{E}_j \quad \text{or} \quad \mathcal{E}_j = Y_j - f(t_j, \vec{q}_0).$$

Since it is assumed that the error $\mathcal{E}_j$ is *i.i.d.*, a plot of the residuals $r_j = y_j - f(t_j, \hat{q})$ vs. t_j should be random. Also, the error in the constant variance case does not depend on $f(t_j, q_0)$, and so a plot of the residuals $r_j = y_j - f(t_j, \hat{q})$ vs. $f(t_j, \hat{q})$ should also be random. Therefore, *if* the error has constant variance, then a plot of the residuals $r_j = y_j - f(t_j, \hat{q})$ against t_j and against $f(t_j, \hat{q})$ should both be random. If not, then the constant variance assumption is suspect.

We turn next to questions of what to expect if this residual test is applied to a data set that has nonconstant variance (NCV) generated error. That is, we wish to investigate what happens if the data are incorrectly assumed to have CV error when in fact they have NCV error. Since in the NCV example, $R_j = Y_j - f(t_j, \vec{q}_0) = f(t_j, \vec{q}_0) \mathcal{E}_j$ depends upon the deterministic model $f(t_j, \vec{q}_0)$, we should expect that a plot of the residuals $r_j = y_j - f(t_j, \hat{q})$ vs. t_j should exhibit some type of pattern. Also, the residuals actually depend on $f(t_j, \hat{q})$ in the NCV case, and so as $f(t_j, \hat{q})$ increases the variation of the residuals $r_j = y_j - f(t_j, \hat{q})$ should increase as well. Thus $r_j = y_j - f(t_j, \hat{q})$ vs. $f(t_j, \hat{q})$ should have a fan shape in the NCV case.

In summary, if a data set has nonconstant variance generated data, then

$$Y_j = f(t_j, \vec{q}_0) + f(t_j, \vec{q}_0)\,\mathcal{E}_j \quad \text{or} \quad \mathcal{E}_j = \frac{Y_j - f(t_j, \vec{q}_0)}{f(t_j, \vec{q}_0)}.$$

If the distributions of $\mathcal{E}_j$ are *i.i.d.*, then a plot of the *modified residuals* $r_j^m = (y_j - f(t_j, \hat{q}))/f(t_j, \hat{q})$ vs. t_j should be random in nonconstant variance generated data. A plot of $r_j^m = (y_j - f(t_j, \hat{q}))/f(t_j, \hat{q})$ vs. $f(t_j, \hat{q})$ should also be random.

Another question of interest concerns the case in which the data are incorrectly assumed to have nonconstant variance error when in fact they have constant variance error. Since $Y_j - f(t_j, \vec{q}_0) = \mathcal{E}_j$ in the constant variance case, we should expect that a plot of $r_j^m = (y_j - f(t_j, \hat{q}))/f(t_j, \hat{q})$ vs. t_j as well as that for $r_j^m = (y_j - f(t_j, \hat{q}))/f(t_j, \hat{q})$ vs. $f(t_j, \hat{q})$ will possess some distinct pattern.

There are two further issues regarding residual plots. As we shall see by examples, some data sets might have values that are repeated or nearly repeated a large number of times (for example when sampling near an equilibrium for the mathematical model or when sampling a periodic system over many periods). If a certain value is repeated numerous times (e.g., f_{repeat}) then any plot with $f(t_j, \hat{q})$ along the horizontal axis should have a cluster of values along the vertical line $x = f_{\text{repeat}}$. This feature can easily be removed by excluding the data points corresponding to these high frequency values (or simply excluding the corresponding points in the residual plots). Another common technique when plotting against model predictions is to plot against $\log f(t_j, \hat{q})$ instead of $f(t_j, \hat{q})$ itself which has the effect of "stretching out" plots at the ends. Also, note that the model value $f(t_j, \hat{q})$ could possibly be zero or very near zero, in which case the modified residuals $R_j^m = \frac{Y_j - f(t_j, \hat{q})}{f(t_j, \hat{q})}$ would be undefined or extremely large. To remedy this situation one might exclude values very close to zero (in either the plots or in the data themselves). We chose here to reduce the data sets (although this sometimes could lead to a deterioration in the estimation results obtained). In our examples below, estimates obtained using a truncated data set will be denoted by $\hat{q}_{\text{OLS}}^{\text{tcv}}$ for constant variance data and $\hat{q}_{\text{OLS}}^{\text{tncv}}$ for nonconstant variance data.

3.4.2 An Example Using Residual Plots

We illustrate residual plot techniques by exploring a widely studied model — the logistic population growth model of Verhulst/Pearl [16]

$$\dot{z} = rz\left(1 - \frac{z}{K}\right), \quad z(0) = z_0. \tag{3.94}$$

Here K is the population's carrying capacity, r is the intrinsic growth rate and z_0 is the initial population size. This well-known logistic model describes how populations grow when constrained by resources or competition. We shall discuss this model, its derivation and properties in more detail subsequently in this monograph in Chapter 9. The closed form solution of this simple model is given by

$$z(t) = \frac{K\, z_0 e^{rt}}{K + z_0\,(e^{rt} - 1)}. \tag{3.95}$$

The left plot in Figure 9.2 depicts the solution of the logistic model for $K = 17.5$, $r = .7$ and $z_0 = 0.1$ for $0 \le t \le 25$. If high frequency repeated or nearly repeated values (i.e., near the initial value x_0 or near the the asymptote $x = K$) are removed from the original plot, the resulting truncated plot is given in the right panel of Figure 9.2 (there are no near zero values for this function).

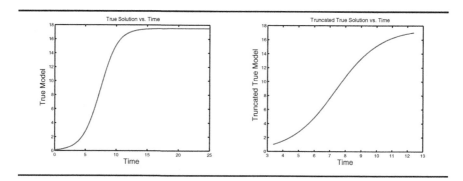

FIGURE 3.4: Original and truncated logistic curve with $K = 17.5$, $r = .7$ and $z_0 = .1$.

For this example we generated both CV and NCV noisy data (we sampled from $\mathcal{N}(0,1)$ random variables to obtain realizations of $\mathcal{E}_j$) and obtained estimates $\hat{q}$ of $\vec{q}_0 = (K, r, z_0)$ by applying either the OLS or GLS method to a realization $\{y_j\}_{j=1}^n$ of the random process $\{Y_j\}_{j=1}^n$. The initial guesses $\vec{q}_{init} = \hat{q}^{(0)}$ along with estimates for each method and error structure are given in Tables 3.1 – 3.4. As expected, both methods do a good job of estimating $\vec{q}_0$, however the error structure was not always correctly specified since incorrect asymptotic formulas were used in some cases.

When the OLS method was applied to nonconstant variance data and the GLS method was applied to constant variance data, the residual plots given below do reveal that the error structure was misspecified. For instance, the plot of the residuals for $\hat{q}_{\mathrm{OLS}}^{\mathrm{ncv}}$ given in Figures 3.7 and 3.8 reveal a fan shaped pattern, which indicates the constant variance assumption is suspect. In addition, the plot of the residuals for $\hat{q}_{\mathrm{GLS}}^{\mathrm{cv}}$ given in Figures 3.9 and 3.10 reveal an inverted fan shaped pattern, which indicates the nonconstant variance as-

TABLE 3.1: Estimation using the OLS procedure with CV data for $\eta = 5$.

$\vec{q}_{\text{init}}$	$\vec{q}_0$	$\hat{q}_{\text{OLS}}^{\text{cv}}$	$\text{SE}(\hat{q}_{\text{OLS}}^{\text{cv}})$	$\hat{q}_{\text{OLS}}^{\text{tcv}}$	$\text{SE}(\hat{q}_{\text{OLS}}^{\text{tcv}})$
17	17.5	1.7500e+001	1.5800e-003	1.7494e+001	6.4215e-003
.8	.7	7.0018e-001	4.2841e-004	7.0062e-001	6.5796e-004
1.2	.1	9.9958e-002	3.1483e-004	9.9702e-002	4.3898e-004

TABLE 3.2: Estimation using the GLS procedure with CV data for $\eta = 5$.

$\vec{q}_{\text{init}}$	$\vec{q}_0$	$\hat{q}_{\text{GLS}}^{\text{cv}}$	$\text{SE}(\hat{q}_{\text{GLS}}^{\text{cv}})$	$\hat{q}_{\text{GLS}}^{\text{tcv}}$	$\text{SE}(\hat{q}_{\text{GLS}}^{\text{tcv}})$
17	17.5	1.7500e+001	1.3824e-004	1.7494e+001	9.1213e-005
.8	.7	7.0021e-001	7.8139e-005	7.0060e-001	1.6009e-005
1.2	.1	9.9938e-002	6.6068e-005	9.9718e-002	1.2130e-005

TABLE 3.3: Estimation using the OLS procedure with NCV data for $\eta = 5$.

$\vec{q}_{\text{init}}$	$\vec{q}_0$	$\hat{q}_{\text{OLS}}^{\text{ncv}}$	$\text{SE}(\hat{q}_{\text{OLS}}^{\text{ncv}})$	$\hat{q}_{\text{OLS}}^{\text{tncv}}$	$\text{SE}(\hat{q}_{\text{OLS}}^{\text{tncv}})$
17	17.5	1.7499e+001	2.2678e-002	1.7411e+001	7.1584e-002
.8	.7	7.0192e-001	6.1770e-003	7.0955e-001	7.6039e-003
1.2	.1	9.9496e-002	4.5115e-003	9.4967e-002	4.8295e-003

TABLE 3.4: Estimation using the GLS procedure with NCV data for $\eta = 5$.

$\vec{q}_{\text{init}}$	$\vec{q}_0$	$\hat{q}_{\text{GLS}}^{\text{ncv}}$	$\text{SE}(\hat{q}_{\text{GLS}}^{\text{ncv}})$	$\hat{q}_{\text{GLS}}^{\text{tncv}}$	$\text{SE}(\hat{q}_{\text{GLS}}^{\text{tncv}})$
17	17.5	1.7498e+001	9.4366e-005	1.7411e+001	3.1271e-004
.8	.7	7.0217e-001	5.3616e-005	7.0959e-001	5.7181e-005
1.2	.1	9.9314e-002	4.4976e-005	9.4944e-002	4.1205e-005

sumption is suspect. As expected, when the correct error structure is specified, the *i.i.d.* test and the model dependence test each display a random pattern (Figures 3.5, 3.6 and Figures 3.11, 3.12).

Also, included in the right panel of Figures 3.5 – 3.12 are the residual plots with the truncated data sets. In those plots only model values between one and seventeen were considered (i.e., $1 \leq y_j \leq 17$). Doing so removed the dense vertical lines in the plots with $f(t_j, \hat{q})$ along the x-axis. Nonetheless, the conclusions regarding the error structure remain the same.

In addition to the residual plots, we can also compare the standard errors obtained for each simulation. At a quick glance of Tables 3.1 - 3.4, the standard error of the parameter K in the truncated data set is larger than the standard error of K in the original data set. This behavior is expected. If we remove the "flat" region in the logistic curve, we actually discard measurements with high information content about the carrying capacity K — see

[4]. Doing so reduces the quality of the estimator for K. Another interesting observation is that the standard errors of the GLS estimate are more optimistic than that of the OLS estimate, even when the non-constant variance assumption is wrong. This example further solidifies the conclusion that before one reports an estimate and corresponding standard errors, there needs to be some assurance that the proper error structure has been specified.

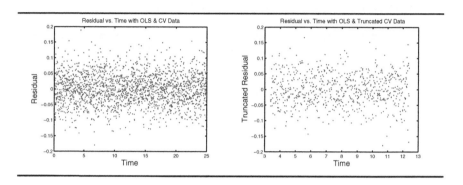

FIGURE 3.5:　Residual vs. time plots: Original and truncated logistic curve for $\hat{q}_{\mathrm{OLS}}^{\mathrm{CV}}$ with $\eta = 5$.

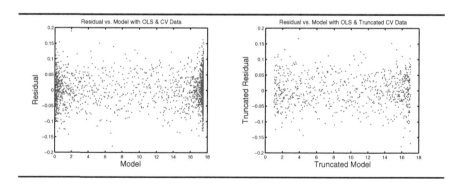

FIGURE 3.6:　Residual vs. model plots: Original and truncated logistic curve for $\hat{q}_{\mathrm{OLS}}^{\mathrm{CV}}$ with $\eta = 5$.

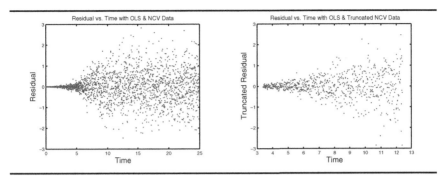

FIGURE 3.7: Residual vs. time plots: Original and truncated logistic curve for $\hat{q}_{\text{OLS}}^{\text{NCV}}$ with $\eta = 5$.

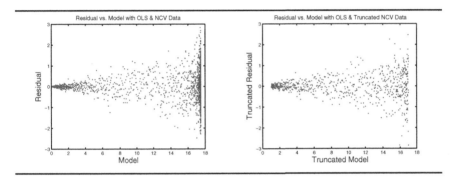

FIGURE 3.8: Residual vs. model plots: Original and truncated logistic curve for $\hat{q}_{\text{OLS}}^{\text{NCV}}$ with $\eta = 5$.

3.5   Statistically Based Model Comparison Techniques

In previous sections we have discussed techniques (e.g., residual plots) for investigating correctness of the assumed *statistical model* underlying the estimation (OLS or GLS) procedures used in inverse problems. To this point we have not discussed correctness issues related to choice of the *mathematical model*. However there are a number of ways in which questions related to the mathematical model may arise. In general, modeling studies [7, 8] can raise questions as to whether a mathematical model can be improved by *more detail* and/or *further refinement*. For example, one might ask whether one can improve the mathematical model by assuming more *detail* in a given mechanism (constant rate vs. time or spatially dependent rate — e.g., see [1] for questions related to time dependent mortality rates during sub-lethal damage in insect

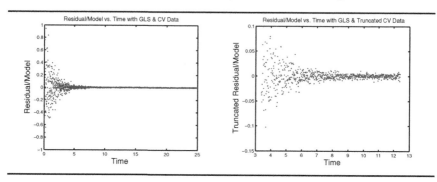

FIGURE 3.9: Residual vs. time plots: Original and truncated logistic curve for $\hat{q}_{\mathrm{GLS}}^{\mathrm{CV}}$ with $\eta = 5$.

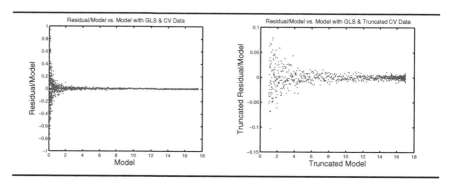

FIGURE 3.10: Modified residual vs. model plots: Original and truncated logistic curve for $\hat{q}_{\mathrm{GLS}}^{\mathrm{CV}}$ with $\eta = 5$.

populations exposed to various levels of pesticides). Or one might question whether an *additional mechanism* in the model might produce a better fit to data — see [5, 6, 7] for *diffusion alone* or *diffusion plus convection* in cat brain transport in grey vs. white matter considerations.

Before continuing, an important point must be made: In model comparison results outlined below, there are really <u>*two models*</u> being compared: the *mathematical model* and the *statistical model*. If one embeds the mathematical model in the *wrong statistical model* (for example, assuming constant variance when this really isn't true), then the mathematical model comparison results using the techniques presented here will be *invalid* (i.e., *worthless*). An important remark in all this is that one must have the mathematical model one wants to simplify or improve (e.g., test whether $\mathcal{V} = 0$ or not in the example below) embedded in the *correct statistical model* (determined in large part by the observation process), so that the comparison actually is *only with regard to the mathematical model*.

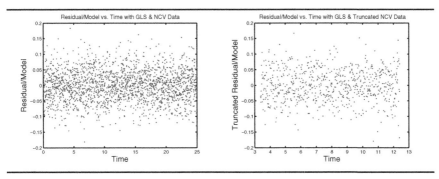

FIGURE 3.11: Modified residual vs. time plots: Original and truncated logistic curve for $\hat{q}_{\mathrm{GLS}}^{\mathrm{NCV}}$ with $\eta = 5$.

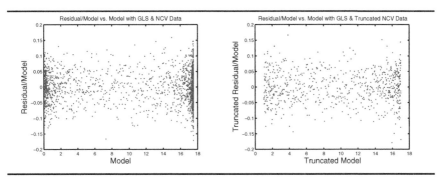

FIGURE 3.12: Modified residual vs. model plots: Original and truncated logistic curve for $\hat{q}_{\mathrm{GLS}}^{\mathrm{NCV}}$ with $\eta = 5$.

To provide specific motivation, we illustrate the formulation of hypothesis testing by considering a mathematical model for a diffusion-convection process. This model was proposed for use with experiments designed to study substance (labelled sucrose) transport in cat brains, which are heterogeneous, containing grey and white matter [7]. In general, the transport of substance in cat's brains can be described by a PDE describing *change in time and space*. This convection/diffusion model, which is widely discussed in the applied mathematics and engineering literature, has the form

$$\frac{\partial u}{\partial t} + \mathcal{V}\frac{\partial u}{\partial x} = \mathcal{D}\frac{\partial^2 u}{\partial x^2}. \tag{3.96}$$

Here, the parameter $\vec{q} = (\mathcal{D}, \mathcal{V})$, which belongs to some admissible parameter set $\mathcal{Q}$, denotes the diffusion coefficient $\mathcal{D}$ and the bulk velocity $\mathcal{V}$ of the fluid, respectively. Our problem: test whether the parameter $\mathcal{V}$ plays a significant role in the mathematical model. That is, if the model (3.96) represents a diffusion-convection process, we seek to determine whether diffusion alone or diffusion plus convection best describes transport phenomena represented in cat brain data sets $\{y_{ij}\}$ for $\{u(t_i, x_j; \vec{q})\}$, the concentration of labelled sucrose at times $\{t_i\}$ and location $\{x_j\}$. We thus might wish to test the null hypothesis H_0 that diffusion alone best describes the data versus the alternative hypothesis H_A that convection is also needed. We then may take $H_0 : \mathcal{V} = 0$ and the alternative $H_A : \mathcal{V} \neq 0$. Consequently, the restricted parameter set $\mathcal{Q}_H \subset \mathcal{Q}$ defined by

$$\mathcal{Q}_H = \{\vec{q} \in \mathcal{Q} : \mathcal{V} = 0\}$$

will be important. To carry out these determinations, we will need some model comparison tests of analysis of variance (ANOVA) type [14] from statistics involving residual sum of squares (RSS) in least squares problems.

3.5.1   RSS Based Statistical Tests

In general, we assume an inverse problem with mathematical model $f(t, \vec{q})$ and n observations $\vec{Y} = \{Y_j\}_{j=1}^n$. We define an OLS performance criterion

$$J_n(\vec{q}) = J_n(\vec{Y}, \vec{q}) = \frac{1}{n} \sum_{j=1}^n [Y_j - f(t_j, \vec{q})]^2,$$

where our *statistical model* again has the form

$$Y_j = f(t_j, \vec{q}_0) + \mathcal{E}_j, \quad j = 1, \dots, n,$$

with $\{\mathcal{E}_j\}_{j=1}^n$ being independent and identically distributed, $E(\mathcal{E}_j) = 0$ and constant variance $\mathrm{var}(\mathcal{E}_j) = \sigma^2$. As usual $\vec{q}_0$ is the "true" value of $\vec{q}$ which we assume to exist. As noted above, we use $\mathcal{Q}$ to represent the set of all the admissible parameters $\vec{q}$ and assume that $\mathcal{Q}$ is a compact subset of Euclidean space of R^p with $\vec{q}_0 \in \mathcal{Q}$.

Let $q^n(\vec{Y}) = q_{OLS}^n(\vec{Y})$ be the OLS *estimator* using J_n with corresponding *estimate* $\hat{q}^n = q_{OLS}^n(\vec{y})$ for a realization $\vec{y} = \{y_j\}$. That is,

$$q^n(\vec{Y}) = \arg\min_{\vec{q} \in \mathcal{Q}} J_n(\vec{Y}, \vec{q}) \quad \text{and} \quad \hat{q}^n = \arg\min_{\vec{q} \in \mathcal{Q}} J_n(\vec{y}, \vec{q}).$$

We remark that in most calculations, one actually uses an approximation f^N to f, often a numerical solution to the ODE or PDE for modeling the dynamical system. Here we tacitly assume f^N will converge to f as the approximation improves. There are also questions related to approximations of the set $\mathcal{Q}$ when it is infinite dimensional (e.g., in the case of function space parameters such as time or spatially dependent parameters) by finite dimensional discretizations $\mathcal{Q}^M$. For extensive discussions related to these questions, see [8] as well as [6] where related assumptions on convergences

$f^N \to f$ and $Q^M \to Q$ are given. We shall ignore these issues in our presentations, keeping in mind that these approximations will also be of importance in the methodology discussed below in most practical uses.

In many instances, including the motivating example given above, one is interested in using data to address the question whether or not the "true" parameter $\vec{q}_0$ can be found in a subset $Q_H \subset Q$, which we assume for discussions here is defined by

$$Q_H = \{\vec{q} \in Q | H\vec{q} = c\}, \tag{3.97}$$

where H is an $r \times p$ matrix of full rank, and c is a known constant vector. In this case we want to test the *null hypothesis* H_0: $\vec{q}_0 \in Q_H$.

Define then

$$q_H^n(\vec{Y}) = \arg\min_{\vec{q} \in Q_H} J_n(\vec{Y}, \vec{q}) \quad \text{and} \quad \hat{q}_H^n = \arg\min_{\vec{q} \in Q_H} J_n(\vec{y}, \vec{q})$$

and observe that $J_n(\vec{Y}, \hat{q}_H^n) \geq J_n(\vec{Y}, \hat{q}^n)$. We define the related non-negative test statistics and their realizations, respectively, by

$$T_n(\vec{Y}) = n(J_n(\vec{Y}, q_H^n) - J_n(\vec{Y}, q^n))$$

and

$$\hat{T}_n = T_n(\vec{y}) = n(J_n(\vec{y}, \hat{q}_H^n) - J_n(\vec{y}, \hat{q}^n)).$$

One can establish asymptotic convergence results for the test statistics $T_n(\vec{Y})$, as given in detail in [6]. These results can, in turn, be used to establish a fundamental result about more useful statistics for model comparison. We define these statistics by

$$U_n(\vec{Y}) = \frac{T_n(\vec{Y})}{J_n(\vec{Y}, q_n)}, \tag{3.98}$$

with corresponding realizations $\hat{U}_n = U_n(\vec{y})$. We then have the asymptotic result that is the basis of our ANOVA–type tests.

Under reasonable assumptions (very similar to those required in the asymptotic sampling distribution theory discussed in previous sections (see [6, 8, 12, 17])) involving regularity and the manner in which samples are taken, one can prove a number of convergence results including:

(i) The estimators q^n converge to $\vec{q}_0$ with probability one as $n \to \infty$;

(ii) If H_0 is true, U_n converges in distribution to $U(r)$ as $n \to \infty$ where $U \sim \chi^2(r)$, a χ^2 distribution with r degrees of freedom, where r is the number of constraints specified by the matrix H.

Recall that H is the $r \times p$ matrix of full rank defining Q_H and that random variables *converge in distribution* if their corresponding cumulative distribution functions converge point wise at all points of continuity of the limit cdf.

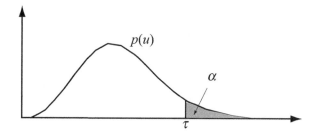

FIGURE 3.13: Example of $U \sim \chi^2(4)$ density.

An example of the χ^2 density is depicted in Figure 3.13 where the density for $\chi^2(4)$ (χ^2 with $r = 4$ degrees of freedom) is graphed. In this figure two parameters (τ, α) of interest are shown. For a given value τ, the value α is simply the probability that the random variable U will take on a value greater than α. That is, $P(U > \tau) = \alpha$ where in hypothesis testing, α is the *significance level* and τ is the *threshold*.

We wish to use this distribution to test the null hypothesis, H_0, which we approximate by $U_n \sim \chi^2(r)$. If the test statistic, $\hat{U}_n > \tau$, then we *reject H_0* as false with confidence level $(1 - \alpha)100\%$. Otherwise, we *do not reject H_0* as true. We emphasize that care should be taken in stating conclusions: we either reject or do not reject H_0 at the specified level of confidence. For the cat brain problem, we use a $\chi^2(1)$ table, which can be found in any elementary statistics text or online and is given here for illustrative purposes; see Table 3.5.

TABLE 3.5: $\chi^2(1)$ values.

α	τ	confidence
.25	1.32	75%
.1	2.71	90%
.05	3.84	95%
.01	6.63	99%
.001	10.83	99.9%

3.5.1.1 P-Values

The minimum value α^* of α at which H_0 can be rejected is called the *p-value*. Thus, the smaller the p-value, the stronger the evidence in the data in support of rejecting the null hypothesis and including the term in the model, i.e., the more likely the term should be in the model. We implement this as follows: Once we compute $\hat{U}_n = \bar{\tau}$, then $p = \alpha^*$ is the value that corresponds to $\bar{\tau}$ on a χ^2 graph and so we reject the null hypothesis at any confidence level c, such that $c < 1 - \alpha^*$. For example, if for a computed $\bar{\tau}$ we find $p = \alpha^* = .0182$, then we would reject H_0 at confidence level $(1 - \alpha^*)100\% = 98.18\%$ or lower. For more information, the reader can consult ANOVA discussions in any good statistics book.

3.5.1.2 Alternative Statement

To test the null hypothesis H_0, we choose a significance level α and use χ^2 tables to obtain the corresponding threshold $\tau = \tau(\alpha)$ so that $P(\chi^2(r) > \tau) = \alpha$. We next compute $\hat{U}_n = \bar{\tau}$ and compare it to τ. If $\hat{U}_n > \tau$, then we *reject* H_0 as false; otherwise, we do not reject the null hypothesis H_0.

3.5.2 Application: Cat-Brain Diffusion/Convection Problem

We summarize use of the model comparison techniques outlined above by returning to the cat brain example discussed in detail in [7, 8]. There were *3 sets of experimental data* examined, under the null-hypothesis $H_0 : \mathcal{V} = 0$.

For *Data Set 1*, we found after carrying out the inverse problems over $\mathcal{Q}$ and $\mathcal{Q}_H$, respectively,

$$J_n(\hat{q}^n) = 106.15 \quad \text{and} \quad J_n(\hat{q}_H^n) = 180.1.$$

In this case $\hat{U}_n = 5.579$ (note that $n = 8 \neq \infty$), for which $p = \alpha^* = .0182$. Thus, we reject H_0 in this case at *any* confidence level less than 98.18%. Thus, we should *reject* that $\mathcal{V} = 0$, which suggests convection is important in describing this data set.

For *Data Set 2*, we found

$$J_n(\hat{q}^n) = 14.68 \quad \text{and} \quad J_n(\hat{q}_H^n) = 15.35,$$

and thus, in this case, we have $\hat{U}_n = .365$, which implies we *do not reject* H_0 with *high degrees of confidence* (p-value very high). This suggests $\mathcal{V} = 0$, which is completely opposite to the findings for Data Set 1.

For the final set (*Data Set 3*) we found

$$J_n(\hat{q}^n) = 7.8 \quad \text{and} \quad J_n(\hat{q}_H^n) = 146.71,$$

which yields in this case, $\hat{U}_n = 15.28$. This, as in the case of the first data set, suggests (with $p < .001$) that $\mathcal{V} \neq 0$ is important in modeling the data.

The difference in conclusions between the first and last sets and that of the second set is interesting and perhaps at first puzzling. However, when discussed with the doctors who provided the data, it was discovered that the first and last set were taken from the *white matter* of the brain, while the other was taken from the *grey matter*. This later finding was consistent with observed microscopic tests on the various matter (micro channels in white matter that promote convective "flow"). Thus, it can be suggested with a reasonably high degree of confidence, that white matter exhibits convective transport, while grey matter does not.

Exercise: Solutions to the MLE

Use the second derivative test to verify that the expressions in equations (3.48) and (3.49) for q_{MLE} and σ^2_{MLE}, respectively, do indeed maximize (3.46).

Project: Statistical Analysis in Inverse Problems Using Simulated Data

The aim of this project is to apply the statistical analysis for inverse problems to the exercise described in Chapter 2. In particular, we use the harmonic oscillator (mass-spring-dashpot) model given by

$$m\frac{d^2y(t)}{dt^2} + c\frac{dy(t)}{dt} + ky(t) = 0$$

with initial conditions

$$y(t_0) = y_0, \qquad \frac{dy(t_0)}{dt} = v_0,$$

or

$$\frac{d^2y(t)}{dt^2} + C\frac{dy(t)}{dt} + Ky(t) = 0$$

with initial conditions

$$y(t_0) = y_0, \qquad \frac{dy(t_0)}{dt} = v_0.$$

The above two models are equivalent when $C = c/m$ and $K = k/m$ if $m \neq 0$. In general, the coefficients C and K are unknown parameters. These parameters can be estimated via a nonlinear least squares estimation problem. Specifically, one seeks $\vec{q} = (C, K)$ to minimize the cost function

$$J(\vec{q}) = \sum_{i=1}^{n} \left| y_m(t_i; \vec{q}) - y_i^d \right|^2,$$

where $y_m(t_i; \vec{q})$ is the model solution to the spring mass dashpot model at time t_i for $i = 1, 2, \ldots, n$, given the parameter set $\vec{q}$ and y_i^d is the data (displacement) collected also at time t_i. In this exercise, we will create *"simulated"* data to be used for estimating the unknown parameters $\vec{q} = (C, K)$. For this, we assume that displacement is sampled at equally spaced time intervals. We will subdivide the time interval $[0, 5]$ into n equal subintervals of length $h = 5/n$. Let y_i^d denote the displacement sampled at time $t_i = ih$, $i = 1, \ldots, N$. For this, use the solution $y(t_i)$ to the spring-mass-dashpot system corresponding to $C = 1$, and $K = 1.5$ and add to each simulated data an error term as follow:

$$\hat{y}_d(t_i) = y_d(t_i) + nl \cdot rand_i,$$

where $rand_i$ are the normally distributed random numbers with zero mean and variance 1.0. Use the MATLAB routine randn to generate an n-vector with random entries. Here, nl is a noise level constant.
    For each of the values $nl = 0.01$, $nl = 0.02$, $nl = 0.05$, $nl = 0.1$, $nl = 0.2$,

1). Estimate the parameters C, and K using the ordinary least squares method.

2). Compute the standard error for each parameter.

3). Report the covariance matrix and discuss the off-diagonal elements.

4). Compare your computed values with "true" values (which we can compute because we know the "true" standard deviation).

Project: Hypothesis Testing Using Experimental Data

The aim of this project is to use the hypothesis testing to investigate phenomena represented in experimental data. In particular, we use the harmonic oscillator (mass-spring-dashpot) model to describe vibrational data from a cantilever beam. The mathematical model is given by

$$m\frac{d^2y(t)}{dt^2} + c\frac{dy(t)}{dt} + ky(t) = 0$$

with initial conditions

$$y(t_0) = y_0, \qquad \frac{dy(t_0)}{dt} = v_0,$$

or

$$\frac{d^2y(t)}{dt^2} + C\frac{dy(t)}{dt} + Ky(t) = 0$$

with initial conditions

$$y(t_0) = y_0, \qquad \frac{dy(t_0)}{dt} = v_0.$$

The above two models are equivalent when $C = c/m$ and $K = k/m$ if $m \neq 0$.

1.) Data collection:

Excite the cantilever beam with a sinusoidal input at the first fundamental frequency of the beam (approximately 6.5 Hz). Use the piezoceramic patches for the exciting actuator. If the measured data is the
displacement (e.g., by using a proximity probe), you would terminate
the exciting input to the beam at $t = t_0$ when $\frac{d}{dt} y(t_0) = 0$ (see Figure
3.14) and measure $\hat{y}_0 = y(t_0)$ (so y_0 and and v_0 are assumed to be given
by observations).

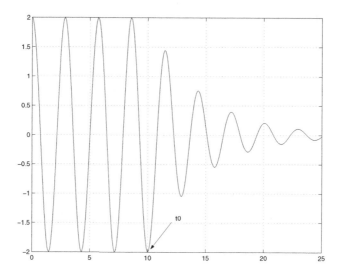

FIGURE 3.14: Beam excitation.

However, for this exercise you will take data on $[t_0, t_1]$: $\hat{a}_i^d$ which are
observations for $\ddot{y}_{\mathrm{mod}}(t_i; q)$ (using an accelerometer). Here q are unknown parameters in the model. It is noted that, since the measured
data are the acceleration, the initial displacement and velocity, y_0 and
and v_0, are indeed unknown (in addition to C and K).

2.) Formulate and carry out the corresponding inverse problem for $q = (C, K, y_0, v_0)$ with

$$J_n(q) = \sum_{i=1}^{n} \left| \hat{a}_i^d - \ddot{y}_{\mathrm{mod}}(t_i; q) \right|^2$$

(a) Estimate $q^* = (C^*, K^*, y_0^*, v_0^*)$ from the data, obtaining $q_n^* = (C_n^*, K_n^*, y_{0_n}^*, v_{0_n}^*)$ so that $J_n(q_n^*)$ is the residual.

(b) Estimate $q^{**} = (0, K^{**}, y_0^{**}, v_0^{**})$ - the undamped model - obtaining $q_n^{**} = (0, K_n^{**}, y_{0_n}^{**}, v_{0_n}^{**})$ so that $J_n(q_n^{**})$ is the residual.

(c) For parts (a) and (b), compute the covariance matrix, standard errors and confidence intervals.

(d) Use the $\chi^2(1)$ test to test for the significance of your improved fit to the data by allowing nontrivial damping $C \neq 0$ in the model. Compute the associated p-value.

3.) Repeat 1.) and 2.) above by exciting the beam with a sinusoidal input at the second fundamental frequency of the beam (approximately 55 Hz). In this case, because the solution is highly oscillatory, it might be difficult to obtain the right initial guesses for the unknown parameters to start the optimization process. One approach to overcome this difficulty is to consider the problem in the frequency domain by considering the fast Fourier transform (fft) of the data and the model solution. One then modifies the parameters C and K so that the frequencies of the solution and data are close. Next, one modifies the initial conditions y_0 and v_0 so that the magnitudes of the fft of the solution and the data are also similar. Now one can use these as initial guesses to carry out the inverse problem in the time domain.

References

[1] H. T. Banks, J.E. Banks, L.K. Dick and J.D. Stark, Estimation of dynamic rate parameters in insect populations undergoing sublethal exposure to pesticides, CRSC-TR05-22, May, 2005; *Bulletin of Mathematical Biology*, **69**, 2007, pp. 2139–2180.

[2] H.T. Banks, M. Davidian, J.R. Samuels, Jr., and K.L. Sutton, An Inverse Problem Statistical Methodology Summary, *CRSC-TR08-01*, January, 2008; Chapter XX in *Statistical Estimation Approaches in Epidemiology*, (edited by Gerardo Chowell, Mac Hyman, Nick Hengartner, Luis M.A Bettencourt and Carlos Castillo-Chavez), Springer, Berlin Heidelberg New York, to appear.

[3] H.T. Banks, S. Dediu and S.E. Ernstberger, Sensitivity functions and their uses in inverse problems, *J. Inverse and Ill-posed Problems*, **15**, 2007, pp. 683–708.

[4] H.T. Banks, S.L. Ernstberger and S.L. Grove, Standard errors and confidence intervals in inverse problems: Sensitivity and associated pitfalls, CRSC-TR06-10, March, 2006; *J. Inverse and Ill-posed Problems*, **15**, 2006, pp. 1–18.

[5] H. T. Banks and B. G. Fitzpatrick, Inverse problems for distributed systems: statistical tests and ANOVA, LCDS/CCS Rep. 88-16, July, 1988, Brown University; *Proc. International Symposium on Math. Approaches to Envir. and Ecol. Problems*, Springer Lecture Note in Biomath., **81**, 1989, pp. 262–273.

[6] H. T. Banks and B. G. Fitzpatrick, Statistical methods for model comparison in parameter estimation problems for distributed systems, CAMS Tech. Rep. 89-4, September, 1989, University of Southern California; *J. Math. Biol.*, **28**, 1990, pp. 501–527.

[7] H.T. Banks and P. Kareiva, Parameter estimation techniques for transport equations with application to population dispersal and tissue bulk flow models, *J. Math. Biol.*, **17**, 1983, pp. 253–272.

[8] H.T. Banks and K. Kunisch, *Estimation Techniques for Distributed Parameter Systems*, Birkhäuser, Boston, 1989.

[9] R.J. Carroll and D. Ruppert, *Transformation and Weighting in Regression*, Chapman & Hall, New York, 1988.

[10] G. Casella and R. L. Berger, *Statistical Inference,* Duxbury, California, 2002.

[11] M. Davidian and D. Giltinan, *Nonlinear Models for Repeated Measurement Data,* Chapman & Hall, London, 1998.

[12] B. Fitzpatrick, *Statistical Methods in Parameter Identification and Model Selection,* Ph.D. Thesis, Division of Applied Mathematics, Brown University, Providence, RI, 1988.

[13] A. R. Gallant, *Nonlinear Statistical Models,* Wiley, New York, 1987.

[14] F. Graybill, *Theory and Application of the Linear Model,* Duxbury, North Scituate, MA, 1976.

[15] R. I. Jennrich, Asymptotic properties of non-linear least squares estimators, *Ann. Math. Statist.,* **40**, 1969, pp. 633–643.

[16] M. Kot, *Elements of Mathematical Ecology,* Cambridge University Press, Cambridge, 2001.

[17] G.A.F. Seber and C.J. Wild, *Nonlinear Regression,* J. Wiley & Sons, Hoboken, NJ, 2003.

Chapter 4

Mass Balance and Mass Transport

4.1 Introduction

Mass transfer is important in many areas of science and engineering. Many familiar phenomena involve mass transfer:

- The spreading of odorous gas in a room.

- Liquid in an open pail of water evaporating into surrounding air.

- A piece of sugar added to a cup of coffee eventually dissolving by itself into the surrounding solution.

- Transport of chemical substances into the red blood cells.

- Transport of O_2 throughout the human body — systemic and cellular.

The most elementary approach to mass transport is compartmental analysis. Compartmental modeling has been and is being used widely in many branches of biology, biomedicine, and in pharmacokinetics as well as in physical modeling. Indeed, one can find examples of compartmental modeling in almost any publication of the major journals in physiology and pharmacology. In addition, there are several books that cover both the theory and applications of compartmental modeling, e.g., [3, 4], while several books have chapters giving introductions to compartmental analysis as well as its applications (see for instance, [1, 5, 6, 7, 8]).

4.2 Compartmental Concepts

A compartment is an abstraction used often in biological (and other scientific) models. It may of course be a physical entity, a distinct space having discernible boundaries across which material (energy) moves at a measurable rate (and for which, as a rule, an "inside" and "outside" are readily distinguishable). More generally, we might take as a compartment any anatomical, physiological, chemical, or physical subdivision of a system throughout which the behavior (e.g., concentration) of a given substance is uniform. It can also be useful to compartmentalize in terms of different types of molecules or chemical forms (e.g., hemoglobin, red blood cells, blood plasma). We might then make a formal definition of a compartment as follows: if a substance S is present in a system in several distinguishable forms or locations and if S passes from one form or location to another form or location at a measurable rate, then each form or location constitutes a separate *compartment* for S.

The compartment concept represents a system as a set of interconnecting components or subsystems. We further remark that the compartments (subsystems) do not always correspond to physically identifiable components. A couple of very simple examples serve to illustrate this concept. In studying certain diseases, it is convenient to regard each *stage* of the disease as a compartment and to construct a mathematical model based on the transfer between them. Another common example of tracer studies involves red blood cells suspended in an isotonic (uniform tension or osmotic pressure) fluid. In this case one might be interested in the concentration of radioactive potassium ions in the million of separate physical compartments. However, for modeling uptake phenomena, it is most likely that one would consider the collection of red blood cells as a whole and formulate a two-compartment model consisting of a fluid compartment and a red blood cell "compartment."

These examples illustrate the fact that it is the behavior of a substance S in a system which determines the compartmentalization of the system and not necessarily the physical situation itself. Differences in how investigators perceive this "behavior" often lead to the dramatically different compartmentalizations of a given system found in the literature. For an example, one might be surprised at the wide range of models used to describe the glucose homeostatic system in mammals.

In the modeling of mass transport between compartments, several assumptions are commonly made. Among these are:

(i) constant-volume compartments,

(ii) well-mixed compartments, and

(iii) for systems in which transport is across a membrane, constancy of the transport coefficient K (discussed further below) in time.

Whether any or all of these assumptions can be justified depends very much on the nature of the phenomena and systems being modeled. While a decision to posit (i) is usually rather straightforward, support of (ii) is often more difficult. There are a number of major contributors to rapid distribution within a compartment, including

(a) stirring or mixing by currents within the body of the solution,

(b) transportation (convection) by a flowing stream, and

(c) diffusion (thermal motion of solute molecules).

Contributions to well-mixing by (a) and (b) can be valid even when the distances (compartment size) are substantial, while (c) is usually a valid component only in the case of small-volume compartments.

The convenience of this type of decomposition (compartmentalization) is that it leads directly to a set of equations based on simple *balance relations*. This can be stated simply as:

change in compartment j = (sum of all transfers into compartment j)

$\qquad$ −(sum of all transfers out of compartment j)

$\qquad$ +(creation within compartment j)

$\qquad$ −(destruction within compartment j).

4.3 Compartment Modeling

To illustrate the ideas behind the concept of a compartment and how it is used, we discuss the simplest example of a two compartment model. Consider two chambers separated by a membrane with solute S and water in each chamber (see Figure 4.1).

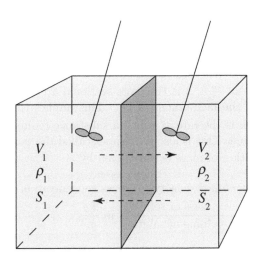

FIGURE 4.1: Two chamber compartments separated by a membrane.

Assume that each chamber is well-mixed (or well-stirred); that is, when the solute S is added to the water it is instantly distributed throughout the chamber. This process is slower for liquid than gas and is slowest for solid (it can be achieved by mixing or by convection by a stream.)

Naturally, one is faced with the following questions. What are the compartments? How many? The answers depend very much on how the solute behaves in the system.

- If the membrane is highly permeable (full of holes), one compartment is adequate to describe the concentration of solute. In this case, equilibration is essentially instantaneous.

- If the membrane is impermeable (no transport across membrane occurs), only one compartment (the one to which solute is added) is needed to model the solute concentration.

- If the membrane is permeable, two compartments are needed and transport of solute between the compartments must be modeled.

In addition, in the modeling of mass transport between compartments separated by a membrane, the above assumptions (i)-(iii) are usually made. The basic parameter involved in membrane separated compartmental exchange is called the *transport coefficient*. It is usually denoted by K and is proportional to physical properties of the compartments. In particular,

$$K \propto \frac{A}{\delta}$$
$$= c\frac{A}{\delta},$$

where c, the proportional constant, is called the membrane permeability coefficient with units of $\frac{m^2}{sec}$, A is the cross-sectional area (in units m^2) and δ is the thickness of the membrane (in m). This implies that K has units of $\frac{m^2}{sec}\frac{m^2}{m} = \frac{m^3}{sec}$, which is the rate at which a substance (volume) is transported across the membrane. This is also sometimes called the *volumetric rate*.

Recall the definition of mass density given by

$$\rho = \text{mass density (or mass concentration)}$$
$$= \frac{\text{mass of solute}}{\text{volume of solution}} \text{ in } \frac{kg}{m^3}.$$

By letting V denote the volume of the compartment into or out of which we are modeling solute transport, we can now write

$$\left\{\begin{array}{c} \text{rate of change} \\ \text{of mass in} \\ \text{compartment} \end{array}\right\} = \left\{\begin{array}{c} \text{volumetric} \\ \text{rate} \end{array}\right\} \times \{\text{mass density}\}.$$

Then, we have

$$\frac{d}{dt}m = K\rho.$$

We next consider using these concepts in a two compartment model such as depicted in Figure 4.1. In this formulation we assume:

- A two-compartment system labeled 1 and 2 with constant volumes V_1 and V_2.

- A solute is present and is transported between compartments across the membrane with transport coefficients $K_{1,2}$ (from 1 to 2) and $K_{2,1}$ (from 2 to 1), which for the moment are not assumed to be equal. The masses of solute in compartments 1 and 2 are denoted by m_1 and m_2 respectively.

Simple mass balance considerations in compartment 1 lead to the following differential equation:

$$\frac{d}{dt}m_1 = \text{(rate of transfer into 1)} - \text{(rate of transfer out of 1)}$$
$$= K_{2,1}\rho_2 - K_{1,2}\rho_1.$$

Similarly, for compartment 2 we obtain

$$\frac{d}{dt}m_2 = K_{1,2}\rho_1 - K_{2,1}\rho_2.$$

We may rewrite this in terms of concentrations (or densities) by using $\rho_i = \frac{m_i}{V_i}$ to obtain

$$\frac{d}{dt}\rho_1 = \frac{1}{V_1}[K_{2,1}\rho_2 - K_{1,2}\rho_1]$$
$$\frac{d}{dt}\rho_2 = \frac{1}{V_2}[K_{1,2}\rho_1 - K_{2,1}\rho_2].$$

From the above calculations, we observe the following important consequences:

- If we assume $K_{1,2} = K_{2,1} = K$, then

$$\frac{dm_1}{dt} = K(\rho_2 - \rho_1)$$
$$\frac{dm_2}{dt} = K(\rho_1 - \rho_2)$$
$$= -\frac{dm_1}{dt}$$

by laws of mass conservation. If $\rho_2 > \rho_1$, then $\frac{dm_1}{dt} > 0$ (that is, the mass of solute in chamber 1 increases due to movement of solute from chamber 2 (high concentration) to chamber 1 (low concentration)).

This type of mass transport is called *passive transport* or *molecular* (membrane) *diffusion*. (It is very similar to the manner in which heat is transported in a rod — one observer holds one end of a rod and when the other end is heated, the part that is held will become hotter even though it is not in direct contact with the heat source; thus, heat is said to be transported (conducted) from high concentration or temperature to low concentration or temperature.)

- In terms of mass *concentrations* (or mass *densities*) we have

$$\frac{d\rho_1}{dt} = \frac{K}{V_1}(\rho_2 - \rho_1)$$
$$\frac{d\rho_2}{dt} = \frac{K}{V_2}(\rho_1 - \rho_2).$$

Note that $\dot{\rho}_1 \neq -\dot{\rho}_2$ unless $V_1 = V_2$. That is, in general, *we do not have concentration balance.*

It is important to note: We have *mass conservation* and *not concentration (or density) conservation.*

There are several advantages as well as disadvantages that arise when using simple compartmental models.

Advantages:

It is relatively straightforward to write down the mass balance equation (input – output relation). In addition, the resulting model is a set of ordinary differential equations (often rather easy to solve analytically or numerically).

Disadvantages:

The solution is assumed to be well-mixed. To see the inherent limitations, we can consider, for example, dropping a blue liquid dye into a bucket of water. The dye will diffuse slowly to other parts of the water. That is, the concentration of the dye is different in different parts of the bucket. Often, one can satisfy the well-mixed assumption by considering very small volumes or by stirring the compartment. However this is not always reasonable. For example, the transport of drug in the liver will have different concentrations through different parts of the liver. The concentration of a drug injected into the systemic blood may have different concentrations in different parts of the blood circulation system.

4.4 General Mass Transport Equations

Recall from compartment analysis, we have

$$\frac{dm_1}{dt} \propto (\rho_2 - \rho_1),$$

that is, the rate of change of mass is proportional to concentration difference. This type of transport process is known as *molecular diffusion*. To illustrate this concept, consider the movement of individual molecules, say A and B, in a fluid as depicted in Figure 4.2.

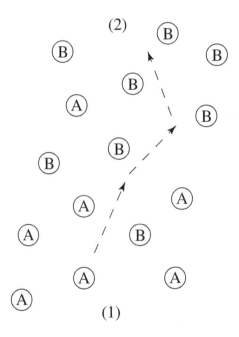

FIGURE 4.2: Binary molecules movement.

Suppose that there are more A molecules near region (1) than near region (2) and since molecules move randomly in both directions, more A molecules will move from (1) to (2) than from (2) to (1). The net transport of A is from a high concentration region to a low concentration region; this is *molecular diffusion*.

We further remark that:

- As molecules move they change directions by bouncing off other molecules after collisions. Since they travel in a random path, molecular diffusion is also called a <u>random walk process</u>.

- To increase the rate of mixing of a substance in solution, the liquid can be mechanically agitated by a device and <u>convective mass transfer</u> will occur (due to movement of the bulk liquid).

Let us now consider a mixture of several species (labeled with index i) in a moving fluid through a pipe as depicted in Figure 4.3.

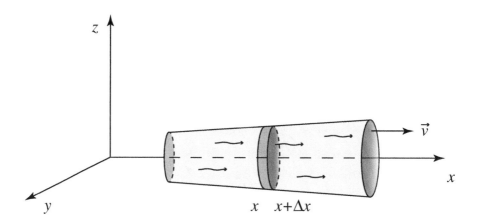

FIGURE 4.3:   Moving fluid through a pipe.

We will formulate a mass balance relationship for species i on a volume element of thickness Δx as shown. The general mass balance on species i is

$$\begin{pmatrix} \text{rate of accumulation} \\ \text{of mass } i \text{ in} \\ \text{volume element} \end{pmatrix} = \begin{pmatrix} \text{rate of} \\ \text{mass } i \text{ entering} \\ \text{face } x \end{pmatrix} - \begin{pmatrix} \text{rate of mass} \\ i \text{ leaving} \\ \text{face } x + \Delta x \end{pmatrix}$$

$$\pm \begin{pmatrix} \text{rate of} \\ \text{generation (or consumption) of} \\ \text{mass } i \text{ (by metabolism or} \\ \text{chemical reaction)} \end{pmatrix}$$

To write down the rate of mass entering and leaving, we need to discuss flux laws for mass transport. (Mass flux is defined as the mass that passes through a unit cross sectional area per unit time.) We do this first in the case in which the carrier fluid itself is stationary, that is, the fluid bulk velocity v is zero.

4.4.1 Mass Flux Law in a Stationary (Non-Moving) Fluid

Since we are dealing, in general, with multiple species, the "concentrations" of the various species may be expressed in numerous ways. We begin by defining mass density (or mass concentration) at a point $p = (x, y, z)$ by

$$\rho(t, x, y, z) = \frac{dm}{dV} = \lim_{\Delta V \to 0} \frac{1}{\Delta V} \int_{\Delta V} m(t, \tilde{x}, \tilde{y}, \tilde{z}) \, dV,$$

where ΔV is a small element of volume containing the point p with $m(t, \tilde{x}, \tilde{y}, \tilde{z})$ being the mass of the particle located at $(\tilde{x}, \tilde{y}, \tilde{z}) \in \Delta V$.

We make the following assumptions for our derivation.

(i) In the small volume element $\Delta V = \Delta x A$ (see Figure 4.4), we have well mixing so that ρ is constant in ΔV.

(ii) Species are uniform in y and z directions (that is, $\rho = \rho(t, x)$).

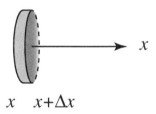

$$x \qquad x+\Delta x$$

FIGURE 4.4: Incremental volume element.

In a diffusing mixture involving multiple species, the various chemical species may be moving at different velocities. Let v_i denote the velocity of species i with respect to a stationary coordinate system. Then we may define the local "mass average velocity" by

$$\bar{v} = \frac{\displaystyle\sum_{i=1}^{n} \rho_i v_i}{\displaystyle\sum_{i=1}^{n} \rho_i}.$$

In some cases one is interested in the velocities of a given species i relative to $\bar{v}$ (or perhaps some other velocity) rather than relative to the stationary coordinate system (v_i). This leads to the definition of the "relative diffusion velocities" v_{ir} given by

$$v_{ir} = v_i - \bar{v} = \text{diffusion velocity of } i \text{ } relative \text{ to } \bar{v}.$$

We may use the mass balance for species i in the element of volume $\Delta V = A\Delta x$ with cross sectional A (which may depend on t and/or x). If we assume no creation or destruction of mass for the present, we may define $\dot{q}_i$ by

$\dot{q}_i$ = rate of mass transport of species i (with mass concentration ρ_i).

We use the compartmental analysis techniques, treating the element of volume ΔV as a "thin membrane" between the immediate "compartments" where the concentrations are $\rho(x)$ and $\rho(x + \Delta x)$, respectively. We find that $\dot{q}_i$ is proportional to $A\frac{\Delta \rho}{\Delta x}$. Then we may write

$$\dot{q}_i = AD_i \frac{\rho_i(x) - \rho_i(x + \Delta x)}{\Delta x},$$

where the constant of proportionality is given by D_i and is called the mass diffusivity constant $\left(\text{in units } \frac{m^2}{sec}\right)$.

Note that here we have assumed that A is approximately constant for the small volume. The above expression is the rate of mass transport in the incremental volume element. To find the rate of mass transport at an arbitrary point x, we let $\Delta x \to 0$ to obtain

$$\dot{q}_i \to -AD_i \frac{\partial \rho_i}{\partial x}.$$

Recall mass flux for species i is $j_i = \dfrac{\text{rate of mass transport}}{\text{cross sectional area}}$. Hence, we have

$$j_i = -D_i \frac{\partial \rho_i}{\partial x} \text{ with units } \frac{kg}{m^2 - sec}. \qquad (4.1)$$

The following remarks are in order:

1. This is known as <u>Fick's first law of diffusion</u> [2], which says that mass flux is proportional to the mass concentration gradient; in general, temperature, pressure gradients, and external forces also affect the flux, but their effects are usually minor and are ignored, or else treated through dependence of the diffusion coefficient D_i on them.

2. We will later see that Fickian diffusion is very similar to Fourier's law of heat conduction and Newton's law of momentum (in one-dimensional problems).

3. The negative sign in (4.1) agrees with the observation that mass flows from high to low mass concentration. If we have $\rho_i(x) < \rho_i(x + \Delta x)$, we find that

$$\frac{\partial \rho_i}{\partial x} > 0$$

and hence net flow is in the opposite direction from the positive x-direction.

4. In three-dimensional problems, these concepts all readily generalize, and for mass density $\rho_i(t, x, y, z)$, we find that the mass flux is given by

$$\vec{j}_i = -D_i \nabla \rho_i.$$

5. The mass flux with respect to the stationary coordinates is given by

$$j_i = \rho_i v_i,$$

and the mass flux with respect to the relative diffusion velocity j_{ir} is given by

$$j_{ir} = \rho_i v_{ir}.$$

4.4.2 Mass Flux in a Moving Fluid

We assume that the bulk velocity is denoted by v, so the total velocity of species relative to the fixed coordinate system is $v_i = v_{i,diff} + v$, and hence the total flux of species i relative to a fixed point in the stationary coordinate system is $j_i^{tot} = \rho_i v_i = \rho_i v_{i,diff} + \rho_i v = j_i^{diff} + j_i^{bulk}$, where we recall the diffusive flux was given by

$$j_i^{diff} = -D_i \frac{\partial \rho_i}{\partial x}.$$

Hence, $j_i^{tot} = -D_i \frac{\partial \rho_i}{\partial x} + \rho_i v$.

Now write the mass balance on a small element:

$$\frac{\partial}{\partial t}[\rho_i \Delta x A(t, x)] = j_i^{tot} A|_x - j_i^{tot} A|_{x+\Delta x} + r_i \Delta x A,$$

where r_i is rate of production (destruction) of species i per unit volume. Dividing by Δx and taking the limit as $\Delta x \to 0$, we obtain

$$\frac{\partial}{\partial t}(\rho_i A) = -\frac{\partial}{\partial x}(j_i^{tot} A) + r_i A$$

$$= -\frac{\partial}{\partial x}\left(-A D_i \frac{\partial \rho_i}{\partial x} + A \rho_i v\right) + r_i A$$

or

$$\frac{\partial}{\partial t}(\rho_i A) + \frac{\partial}{\partial x}(\rho_i v A) = \frac{\partial}{\partial x}\left(A D_i \frac{\partial \rho_i}{\partial x}\right) + r_i A.$$

If A is constant, we obtain the usual mass transport equation

$$\frac{\partial}{\partial t}(\rho_i) + \frac{\partial}{\partial x}(\rho_i v) = \frac{\partial}{\partial x}\left(D_i \frac{\partial \rho_i}{\partial x}\right) + r_i,$$

where the second term is identified with the convective or advective transport, and the third term is diffusive transport.

The above derivation can be generalized to the multiple species case to obtain

$$\frac{\partial}{\partial t}\left(\sum \rho_i A\right) = -\frac{\partial}{\partial x}\left(\sum_i j_i^{tot} A\right) + \sum r_i A,$$

and since $\sum r_i = 0$ (total conservation of mass), $\sum j_i^{tot} = \sum \rho_i v_i^{tot} = \rho v$ (note that the bulk velocity thus agrees with the local mass velocity $v = \frac{\sum \rho_i v_i}{\sum \rho_i}$) and $\sum \rho_i = \rho$, we have

$$\frac{\partial}{\partial t}(\rho) + \frac{\partial}{\partial x}(\rho v) = 0.$$

This is the well known *equation of continuity*.

All of the above generalizes to the three-dimensional problem. In particular, the three-dimensional mass transport equation has the form:

$$\frac{\partial}{\partial t}(\rho_i) + \nabla \cdot (\rho_i v) = \nabla \cdot (D_i \nabla \rho_i) + r_i.$$

The equation of continuity in three-dimensions is given by:

$$\frac{\partial}{\partial t}(\rho) + \nabla \cdot (\rho \vec{v}) = 0. \tag{4.2}$$

If ρ is constant, we obtain $\nabla \cdot \vec{v} = 0$. This is known as *incompressibility* of the fluid in which the solute is contained.

Special cases:

(1) When the bulk velocity, $\vec{v}$, and the reaction rate, r_i, are both zero, we obtain

$$\frac{\partial}{\partial t}(\rho_i) = \nabla \cdot (D_i \nabla \rho_i),$$

which is called *Fick's second law of diffusion* or simply the *diffusion equation*.

(2) In the case the bulk velocity, $\vec{v}$, is zero, we have

$$\frac{\partial}{\partial t}(\rho_i) = \nabla \cdot (D_i \nabla \rho_i) + r_i,$$

which is known as the *reaction-diffusion equation*.

(3) When the diffusion, D_i, is zero, we have

$$\frac{\partial}{\partial t}(\rho_i) + \nabla \cdot (\rho_i \vec{v}) = r_i.$$

This is known as the *plug-flow, ideal tubular*, or *unmixed flow model*. Here the flow of the fluid is orderly with no element of fluid mixing or overtaking (see Figure 4.5). A necessary and sufficient condition for plug flow is that the residence time is the same for each species.

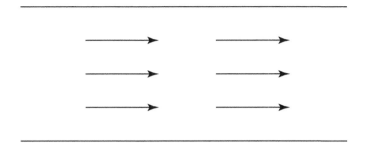

FIGURE 4.5:   Plug flow model.

Exercise: Transport Equations

In the literature one also often finds mass transport in terms of molar concentration c_i and mass fraction w_i. This exercise will provide experience in deriving mass transport equations in terms of these variables.

(i) Define the <u>molar</u> concentration c_i of species i by $c_i = \rho_i/M_i$, where ρ_i is mass concentration (in units kg/m^3) of species i and M_i is the molecular weight (in units kg/moles) so that c_i has units moles of i/m^3. Define the mass fraction $w_i = \rho_i/\rho$, where ρ is the total mass density $\rho = \sum \rho_i$.

Use compartmental analysis to argue that the rate of mass transport at a point x is given by

$$\dot{q}_i = -A\rho D_i \frac{\partial w_i}{\partial x}$$

and the mass flux of species i is given by

$$j_i = \rho D_i \frac{\partial w_i}{\partial x}.$$

Explain when this is equivalent to Fick's first law of diffusion.

(ii) Now use this and mass balance principles to derive the general mass transport equations with diffusive and convective terms in terms of the variable w_i (as opposed to in terms of ρ_i as done earlier in this chapter).

References

[1] J.J. Batzel, F. Kappel, D. Schneditz and H.T. Tran, *Cardiovascular and Respiratory Systems: Modeling, Analysis, and Control*, SIAM, Philadelphia, 2006.

[2] C. J. Geankoplis, *Transport Processes and Unit Operations*, Prentice Hall, Englewood Cliffs, 1993.

[3] K. Godfrey, *Compartmental Models and Their Applications*, Academic Press, New York, 1983.

[4] J.A. Jacquez, *Compartmental Analysis in Biology and Medicine*, The University of Michigan Press, Ann Arbor, 1985.

[5] R.K. Nagle, E.B. Saff and A.D. Snider, *Fundamentals of Differential Equations and Boundary Value Problems*, Pearson Education, Inc., Boston, 2004.

[6] M. Reddy, R.S. Yang, M.E. Andersen and H.J. Clewell, III, *Physiologically Based Pharmacokinetic Modeling: Science and Applications*, Wiley-Interscience, Malden, 2005.

[7] S. Strauss and D.W.A. Bourne, *Mathematical Modeling of Pharmacokinetic Data*, CRC Press, Boca Raton, 1995.

[8] G.G. Walter and M. Contreras, *Compartmental Modeling with Networks*, Birkhäuser, Boston, 1999.

Chapter 5

Heat Conduction

5.1 Motivating Problems

5.1.1 Radio-Frequency Bonding of Adhesives

Radio-frequency (RF) curing of adhesives is a commercially important process which is used in a number of applications. These include the fixation of prosthetic joints in some fields of medicine, the acceleration of adhesive setting in the woodworking industry, and the bonding of parts in the automotive industry. More specifically, in the automobile industry, the use of non-metallic automotive exterior body panels has grown significantly over the last decade. The most common of these materials is sheet molding compound (SMC), a glass-reinforced polyester which provides corrosion resistance, weight reduction, and complex shape molding capability. These parts are typically molded in two layers and adhesively bonded in sandwich fashion around their perimeters to form rigid structures.

The adhesive is commonly applied in a viscous liquid or paste form. Radio frequency, or dielectric, heating is often used to accelerate the cure rate of the adhesive. In this application, the SMC/adhesive/SMC joint is placed between two electrodes (Figure 5.1). These electrodes then make contact with the joint, compressing it to the desired adhesive bonding thickness. A high voltage electric field, oscillating at approximately 30 MHz, then passes through the joint for a predetermined period of time at preset power levels, exciting polar or ionic species in the adhesive materials and generating heat. In comparison to common adhesives, the SMC is dielectrically relatively inactive. Significant heat can be generated within the adhesive, however, causing it to rapidly undergo a phase transition from liquid to solid (curing), and effectively bonding the two substrates to each other. This process, which can be closely simulated on a laboratory scale using a smaller version of the RF bonding equipment described above, provides us with a physically interesting problem. We must deal with thermally dependent nonlinearities arising from the radio-frequency field itself (i.e., temperature dependent input terms as well as conductivities), and complex internal phase transitions which are parametrized by the degree of cure. This process thus provides us with a problem that is mathematically very interesting. It combines serious modeling issues, mathematical analysis, and computational methodology, while providing a foundation for necessary parameter estimation problems and nonlinear control methodology development.

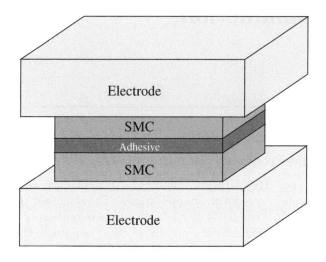

FIGURE 5.1: Diagram of SMC-adhesive-SMC joint.

This industrial problem was a joint collaborative effort between scientists at Lord Corporation (Cary, North Carolina) and faculty and graduate students at North Carolina State University. The goal is to model the radio-frequency curing of epoxy adhesives in bonding of composites. For a detailed development of the mathematical model for the heat transfer through the joint we refer the reader to [1]. The model is a version of the "heat equation" of Fourier fame plus terms that take into account the internal exothermic reaction (which is part of the curing process) as well as the heat generated by the conversion of electrical energy to molecular vibrational energy.

5.1.2 Thermal Testing of Structures

Recently, associated with the use of fiber-reinforced composite materials as well as with more traditional composite metal alloys for aerospace structures, there is growing interest in the detection and characterization of structural flaws (e.g., cracks, delamination, and corrosion) that may not be detectable by visual inspection. An evaluation procedure for such damage detection is of paramount importance in the context of aging aircraft (both civilian and military). One recent effort has focused on nondestructive evaluation (NDE) methods based on the measurement of thermal diffusivity in composite materials (see, e.g., [6]). The idea of this approach is embodied in Figure 5.2.

In [2] the search for structural flaws in materials is formulated as an inverse

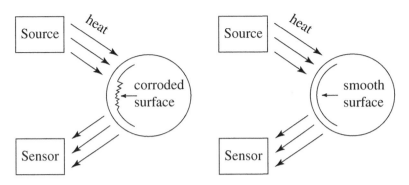

FIGURE 5.2: A schematic diagram of the NDE method for the detection of structural flaws. The sensor measures the surface temperature, and the measured temperature is different for the smooth versus the corroded surface.

problem for a heat diffusion system. From a physical point of view, the system state is the temperature distribution as a function of time and space, the boundary input represents the thermal source (for example, by a laser beam) and the output corresponds to the observation of the temperature distribution at the surface of the material (for example, by an infrared imager); see Figure 5.2 and [6] for more details. The problem is then of identifying, from input and output data, the geometrical structure of the boundary (i.e., the corroded surface). The mathematical model, which relates front surface temperature (the output data) and back surface "geometry," is described by the heat equation with appropriate initial and boundary conditions (see [2] for a detailed description).

5.2 Mathematical Modeling of Heat Transfer

5.2.1 Introduction

In addition to the two examples discussed in §5.1, the transfer of energy in the form of heat occurs in numerous industrial production problems, including those in the chemical industry, the paper industry, and numerous other production processes. For examples, heat transfer occurs in the drying of lumber, chilling of food and biological materials, combustion problems (burning of fuel), and evaporation processes.

In general, heat transfer is energy in transit due to temperature differences and hence "energy balance" is the underlying conservation principle. This transit of energy can occur through conduction, convection, and/or radiation.

- *Conduction.* Conduction generally refers to heat transfer related to *molecular activity* and may be correctly viewed as the transfer of energy from the more energetic to the less energetic particles of a substance or material due to direct interaction between the particles. This type of transfer is present to some extent in all solids, gases, or liquids in which a temperature gradient exists. It is associated with an empirically based rate formulation known as *Fourier's law* to be discussed below. The conduction mode of heat transfer can be related to the random motion of molecules in a gas or substance undergoing no bulk motion or macroscopic movement and is therefore termed diffusion of energy or heat diffusion.

- *Convection.* Heat transfer can also occur in a gas or fluid undergoing bulk or macroscopic motion. The molecules, or aggregates of molecules, move collectively and, in the presence of temperature differences, give rise to energy transfer. The molecules retain, of course, their random motion and thus the energy transfered is a superposition of energy transfer due to random motion of particles as well as due to bulk motion of the fluid. The cumulative transport is usually called *convection* while the transfer due to bulk motion alone is called *advection*, although this clear distinction is not always made. In modeling, a distinction can be made between forced convective heat transfer, where a fluid is forced to flow past a solid surface by a pump, for example, and natural or free convection which arises most often when a gas or fluid passes over a surface when the two are at different temperatures causing a circulation due to a density difference resulting from the temperature differences in the fluid. For either case, the associated empirical rate "law" is called *Newton's law of cooling*.

- *Radiation.* Thermal *radiation* refers to energy emitted by matter at a finite positive temperature. This is usually attributed to changes in electron configurations in atoms and molecules that result in the emission of energy via electromagnetic waves or photons and may occur in solids, fluids or gases. The most important example of radiation is the transport of heat to the earth from the sun. The associated quantitative rate "law" is given by the *Stefan-Boltzmann* law.

5.2.2 Fourier's Law of Heat Conduction

For general molecular transport, all three main types of rate transfer processes — momentum transfer, heat transfer, and mass transfer — are characterized by the same general type of equation. This basic equation is given as follows:

$$\text{rate of a transfer process} = \frac{\text{driving force}}{\text{resistance}}. \tag{5.1}$$

Equation (5.1) simply states that in order to transfer a property (for example, heat) a driving force needs to overcome a resistance.

The transfer of heat by conduction also follows this basic principle and is known as Fourier's law of heat conduction in fluids or solids. It is written mathematically as

$$\dot{q} = -kA\frac{\partial u}{\partial x},$$ (5.2)

where $\dot{q}$ is the rate of heat transfer and is given in units of power, i.e., watts (W), where $1W = 1J/sec = .23885$ calories/sec, A is the cross-sectional area normal to the direction of heat flow in m^2, k is the *thermal conductivity* in W/m · K, u is the temperature in K (or °C), and x is the distance in m. This "law" is based on phenomenological or empirical observations (such as a constitutive assumption or "law" in particle mechanics, i.e., Newton's second "law" of motion, $F = ma$) as opposed to being based on first principles. It satisfies our intuition that the rate of heat transfer across a surface should be proportional to the surface area A and the temperature difference (i.e., the limit of $(u(x + \Delta x) - u(x))/\Delta x$) from one side to the other. The minus sign in (5.2) indicates that heat will "flow" from regions of high temperature to regions of low temperature. In general, the thermal conductivity k (which is a measure of the material's ability to transfer or "conduct" heat) may depend on t, x or even the temperature u. Furthermore, the cross-sectional area A may depend on x (a nonuniform geometry) and/or t (a "pulsating" solid or biological compartment). However, in our fundamental development of the heat equation to be presented in the next section, we shall assume that both k and A are constant (uniform in space and time).

5.2.3 Heat Equation

We begin by considering the unsteady heat transfer problem in one direction in a solid. To derive the conduction equation in one dimension, we refer to Figure 5.3, which depicts a small section of a one-dimensional cylindrical rod centered about an arbitrary point x.

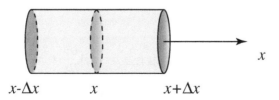

FIGURE 5.3: Transient conduction in one-dimensional cylindrical rod.

We make the following assumptions:

(i) Heat transfer is by conduction;

(ii) Heat transfer is along the x-axis;

(iii) Temperature is uniform over a cross-section;

(iv) We have perfect insulation, hence no heat is escaping from the sides of the cylindrical rod.

Let $u(t, x)$ denote the temperature at x at a given time t and H denote the amount of heat (energy) in units of calories (a calorie is defined as the amount of energy required to raise 1 gm of water 1 °C). Heat may also be given in units of Joules (1 cal = 4.19 J or 1 J = .23885 cal). We expect the amount of heat in an element of mass to be proportional to both the mass and the temperature. This motivates the quantitative expression for *heat*:

$$H = c_p mu, \tag{5.3}$$

where c_p is the specific heat, a constant of proportionality which depends on the material, and m is the mass. The specific heat is given at a constant volume and has units $\dfrac{\text{J}}{\text{kg·K}}$.

We are now ready to turn to energy balance in the small element of the volume between $x - \Delta x$ and $x + \Delta x$ as shown in Figure 5.3. Since the wall of the cylindrical rod is insulated, if we assume that there is no heat generated inside the cylinder, then we have

the net rate of heat accumulation = rate of heat input

$$-\text{rate of heat output.} \tag{5.4}$$

We assume without loss of generality that heat flow is from left to right (i.e., $\frac{\partial u}{\partial x} < 0$). The rate of heat input to the cylinder is

$$\text{rate of heat input} = \dot{q}|_{x - \Delta x} = -kA\frac{\partial u}{\partial x}(t, x - \Delta x). \tag{5.5}$$

Also,

$$\text{rate of heat output} = \dot{q}|_{x + \Delta x} = -kA\frac{\partial u}{\partial x}(t, x + \Delta x). \tag{5.6}$$

The rate of heat accumulation in the elemental volume $2A\Delta x$ is

$$\text{rate of heat accumulation} = \frac{\partial H}{\partial t}, \tag{5.7}$$

and by using the expression (5.3) for heat, we obtain

$$\frac{\partial H}{\partial t} = \frac{\partial}{\partial t}(c_p m u)$$

$$= \frac{\partial}{\partial t}(c_p(2\Delta x \rho A)u(t,x))$$

$$= 2\Delta x c_p \rho A \frac{\partial u(t,x)}{\partial t}, \tag{5.8}$$

where ρ denotes the mass density of the cylindrical rod. Substituting equations (5.5), (5.6), and (5.8) into (5.4) and dividing by $2\Delta x A$, we have

$$\rho c_p \frac{\partial u}{\partial t} = \frac{k\frac{\partial u}{\partial x}(t, x + \Delta x) - k\frac{\partial u}{\partial x}(t, x - \Delta x)}{2\Delta x}.$$

Letting $\Delta x \to 0$ we obtain

$$\rho c_p \frac{\partial u}{\partial t} = \frac{\partial}{\partial x}(k\frac{\partial u}{\partial x}),$$

or, since k is independent of x,

$$\frac{\partial u}{\partial t} = \left(\frac{k}{\rho c_p}\right)\frac{\partial^2 u}{\partial x^2}.$$

This can be written as

$$\frac{\partial u}{\partial t} = \alpha \frac{\partial^2 u}{\partial x^2}, \tag{5.9}$$

where $\alpha \equiv \frac{k}{\rho c_p}$ is the *thermal diffusivity* in $\frac{m^2}{sec}$. Equation (5.9) is known as the *one-dimensional heat equation*. Since k is the material's ability to conduct heat and ρc_p is the volumetric heat capacity (ability of the material to store heat), the thermal diffusivity represents the ability of the material to conduct thermal energy relative to its ability to store it.

A similar derivation when the heat flow is from right to left (i.e., $\frac{\partial u}{\partial x} > 0$) will give the same partial differential equation (5.9) for the heat conduction in a one-dimensional cylindrical rod.

Before turning to the three-dimensional version of the above quantitative description of heat transfer, we note that heat conduction and Fourier's law are often discussed in terms of *heat flux* which is the rate of heat transfer (in the direction x) per unit cross-sectional area and is given by

$$\Phi = \frac{\dot{q}}{A} = -k\frac{\partial u}{\partial x}. \tag{5.10}$$

For heat flux through a general (smooth) surface in three dimensions, the above formula (5.10) is generalized to have the form

$$\Phi = -k\nabla u \cdot \hat{n}, \tag{5.11}$$

where $\hat{n}$ is the unit outward normal vector to the surface ($\hat{n} = \pm \hat{i}$ in the above one-dimensional case).

We now consider a general region V and an arbitrary infinitesimal volume ΔV enclosed by a surface ΔS (see Figure 5.4).

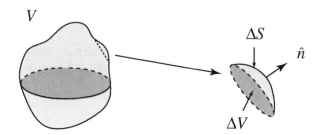

FIGURE 5.4:   (a) A general three-dimensional region. (b) An infinitesimal volume.

We will formulate heat balance equations for the infinitesimal volume ΔV. First, we have

$$\text{rate of heat accumulation in } \Delta V = \frac{\partial H}{\partial t}$$

$$= \frac{\partial}{\partial t} \left(\int_{\Delta V} \rho c_p u \, dV \right)$$

$$= \int_{\Delta V} \rho c_p \frac{\partial u}{\partial t} \, dV. \qquad (5.12)$$

Also, from (5.11), the heat flux across the boundary ΔS of ΔV is given by

$$\Phi = -k\nabla u \cdot \hat{n}.$$

If $\nabla u \cdot \hat{n}$ is positive, we have heat flow *into* the infinitesimal element, so the rate of change of heat across a surface element dS is given by

$$-\Phi dS = k\nabla u \cdot \hat{n} \, dS,$$

which is positive (i.e., the temperature u is *increasing* in the element along the $\hat{n}$ direction.). In this case Φ is negative, i.e., heat is entering the region. If $\nabla u \cdot \hat{n}$ is negative, we have heat flow *out of* the element and the rate of change is again given by

$$-\Phi dS = k\nabla u \cdot \hat{n} \, dS,$$

which is negative (i.e., the temperature u in the element is *decreasing* along the $\hat{n}$ direction). In this case the flux Φ is positive, i.e., heat is leaving the region. In either case, the rate of change of heat in the volume ΔV is given by summing the rate (or the negative of the flux) across the boundary surface area:

$$\frac{\partial H}{\partial t} = \int_{\Delta S} -\Phi \, dS$$
$$= \int_{\Delta S} k \nabla u \cdot \hat{n} \, dS. \tag{5.13}$$

By Gauss' Theorem (the divergence theorem) [8] (see also Appendix B) we find that this last expression (5.13) can be rewritten as

$$\frac{\partial H}{\partial t} = \int_{\Delta V} \nabla \cdot (k \nabla u) \, dV. \tag{5.14}$$

Substituting equation (5.12) for the left side, we have

$$\int_{\Delta V} \left[\rho c_p \frac{\partial u}{\partial t} - \nabla \cdot (k \nabla u) \right] dV = 0$$

for any *arbitrary* element ΔV in V. Since ΔV is arbitrary, it follows that we must have

$$\rho c_p \frac{\partial u}{\partial t} = \nabla \cdot (k \nabla u)$$

in V. For constant thermal conductivity k, this equation is simplified to the following heat equation in three dimensions:

$$\rho c_p \frac{\partial u}{\partial t} = k \nabla \cdot (\nabla u)$$
$$= k \nabla^2 u, \tag{5.15}$$

where $\nabla^2 u = \frac{\partial^2 u}{\partial x^2} + \frac{\partial^2 u}{\partial y^2} + \frac{\partial^2 u}{\partial z^2}$. For additional details on the development of the heat equation, the interested reader may consult [3] or [7].

5.2.4 Boundary Conditions and Initial Conditions

Consider a simple ordinary differential equation (which is the steady-state case of equation (5.9))

$$\frac{d^2 u}{dx^2} = 0.$$

It has infinitely many solutions

$$u(x) = c_1 x + c_2,$$

where c_1 and c_2 are constants. Thus, to find a unique solution even in this simple case, one must impose additional conditions on the problem. To find a unique solution to the heat equation (5.9) or (5.15) we also must impose auxiliary equations. We choose these auxiliary equations to describe the state of our heat flow at time "zero" (the beginning of the experiment, for example) and the state of the flow on the boundary of our region. These equations are called initial conditions and boundary conditions, respectively.

We consider the one-dimensional heat equation (5.9) on a region $(0, L)$

$$\frac{\partial u}{\partial t} = \alpha \frac{\partial^2 u}{\partial x^2}, \qquad 0 < x < L, \ t > 0. \tag{5.16}$$

From equation (5.16), there is one time derivative, which implies that we need one initial condition specifying u for all x at a given time. In this case we do this for $t = 0$ and hence have

$$u(0, x) = \psi(x), \qquad 0 < x < L. \tag{5.17}$$

In addition, there are two spatial derivatives, which imply that we need two boundary conditions specifying u for all t at given values of x. For example, we might specify

$$u(t, 0) = u_1(t) \text{ and } u(t, L) = u_2(t), \tag{5.18}$$

and these are known as *Dirichlet* boundary conditions.

In many cases of interest, boundary conditions might be related to heat flux rather than to temperature. That is, we might specify fixed heat flux at one endpoint or both endpoints of the domain. For example, at $x = L$ we might impose

$$\Phi(t, L) = -k \frac{\partial u}{\partial x}(t, L) = f(t). \tag{5.19}$$

The condition (5.19) is called a *Neumann* boundary condition. A special case of (5.19) is

$$k \frac{\partial u}{\partial x}(t, L) = 0,$$

which means that the endpoint at $x = L$ is *insulated*.

Finally, we might combine both conditions (5.18) and (5.19) to obtain a condition of the form

$$k \frac{\partial u}{\partial x}(t, L) + h u(t, L) = g(t), \tag{5.20}$$

which is known as a *Robin* boundary condition. Here, the parameter h has a physical meaning. It is well known that a hot piece of material will cool faster when air is blown or forced by the object. When the fluid or gas (air) outside the solid surface is forced or when we have natural convective flow, the rate of heat transfer from the solid to the fluid, or vice versa, is given by

$$\dot{q} = h A (u_s - u_f), \tag{5.21}$$

TABLE 5.1:   Range of values of h in Newton cooling.

Mechanism	Range of values of h $(\text{W/m}^2\text{K})$
Still air	2.8-23
Moving air	11.3-55
Moving water	280-17,000
Boiling liquids	1,700-28,000
Condensing steam	5,700-28,000

where $\dot{q}$ is the heat transfer rate in W, A is the area in m^2, u_s is the temperature of the solid surface in K, u_f is the average or bulk temperature of the fluid flowing by in K and h is the *convective heat transfer coefficient* or *Newton cooling constant* in $\text{W/m}^2\text{K}$.

The relation (5.21) is referred to as "Newton's law of cooling." Like other "laws," it is not actually a law but one may think of it as a definition for h based on empirical observations. Since we know that when a fluid flows by a solid surface, there is a thin film, which is almost stationary, adjacent to the solid wall which presents most of the resistance to heat transfer, the parameter h is also often called the *film coefficient* or *film conductance*. In general, h can not be predicted theoretically. It is a function of the system geometry, fluid properties, flow velocity, and, in some cases, the temperature difference. In Table 5.1 some values of h are given for different mechanisms of heat transfer and materials (see, for instance, [4, 5]).

If we next divide both sides of equation (5.21) by A, we obtain the convective heat flux

$$\Phi = \frac{\dot{q}}{A} = h(u_s - u_f). \tag{5.22}$$

So, we might specify the heat flux at one interface between the solid and fluid (or air) as

$$-\Phi(t, L) = h[u_f - u(t, L)],$$

which, after substitution of the form for heat flux (5.10), can be rewritten as

$$k\frac{\partial u}{\partial x}(t, L) + hu(t, L) = hu_f.$$

This equation is of the same form as the Robin boundary condition (5.20) and, hence, the meaning of the constant h as discussed above. We note that, if $u_f > u(t, L)$, heat flows into the solid; otherwise, heat flows out of the solid (cooling).

Before we conclude this section, we will describe another type of boundary condition that occurs in some practical applications. This type of boundary condition is related to the third type of heat transfer mechanism — *radiation heat transfer*. We recall that this is basically an electromagnetic mechanism that allows energy to be transported with the speed of light through space. Since it consists of energy in the form of light waves, it obeys the same laws as does light. That is, it travels in straight lines and is transmitted through vacuum and space. The associated quantitative law is given by the *Stefan-Boltzmann law*, expressed as

$$\dot{q}_{\text{max}} = A\sigma u^4, \tag{5.23}$$

where $\dot{q}_{max}$ is the maximum rate of emitted heat in units W. The parameter σ is the Stefan-Boltzmann constant $(= 5.676 \times 10^{-8} \text{ W/m}^2\text{K}^4)$ and u is the (absolute) temperature (in K) of the emitting surface. A body that achieves this rate in either emission or absorption is called a *perfect radiator* or *black body*, respectively. The actual emitted rate of a general surface is given by a somewhat smaller number

$$\dot{q} = \epsilon A\sigma u^4 \tag{5.24}$$

where $0 \leq \epsilon < 1$, with ϵ called the *emissivity* of the surface or body. When $\varepsilon < 1$, we have a *gray body* or a *gray surface*. Bodies (surfaces) also absorb energy, for example, a black body (a perfect absorber) is defined as one that absorbs all radiant energy and reflects none. If $\dot{q}_{\text{abs}}$, $\dot{q}_{\text{inc}}$ represent the rate of energy absorbed and rate of energy incident, then the surface absorptive property is characterized by a parameter α called the *absorptivity* and is defined by

$$\dot{q}_{\text{abs}} = \alpha \dot{q}_{\text{inc}}.$$

(Unfortunately, the same symbol α often is used in the literature for both the conductive diffusivity and for the absorptivity.) For a *gray* surface defined by $\alpha = \epsilon$, the net rate of heat exchange between a surface and its ambient gas is given by

$$\dot{q}_{\text{net}} = \epsilon A\sigma(u_{\text{sur}}^4 - u_{\text{amb}}^4),$$

where u_{sur}, u_{amb} are the surface and ambient temperatures, respectively. Sometimes this net rate of transfer is written as

$$\dot{q}_{net} = h_r A(u_{\text{sur}} - u_{\text{amb}}),$$

where the *radiative heat transfer coefficient* h_r is defined by

$$h_r \equiv \epsilon\sigma(u_{\text{sur}} + u_{\text{amb}})\left(u_{\text{sur}}^2 + u_{\text{amb}}^2\right).$$

We note that treating h_r as a *constant* essentially *linearizes* the radiation rate equation.

Finally, when radiation heat transfer occurs from the surface of a solid, it is usually accompanied by convective heat transfer unless the solid is in vacuum. The appropriate boundary condition is then given by

$$k\frac{\partial u}{\partial n}\Big|_{\partial\Omega} = h\left[u_{amb} - u\,|_{\partial\Omega}\right] + \epsilon\sigma\left[u_{amb}^4 - u^4\,|_{\partial\Omega}\right],$$

where $\partial\Omega$ is the closed surface enclosing the solid.

5.2.5 Properties of Solutions

There is a great deal of mathematical as well as engineering literature related to the heat equation. Many tools for the analysis of the heat equation exist and we remark briefly on several here.

(a) *Uniqueness.* Consider the classical one-dimensional heat equation

$$\frac{\partial u}{\partial t} = \alpha \frac{\partial^2 u}{\partial x^2}, \tag{5.25}$$

defined on a rectangle domain $\{(t,x) : 0 \le t \le T,\ 0 \le x \le L\}$. Then the solution $u(t,x)$ will achieve its maximum value either initially ($t = 0$) or along the sides at $x = 0$ or $x = L$. This is known as the *Maximum Principle*. The minimum value has the same property and we refer the reader to [10] for the proof of this result.

Let $u(t,x)$ denote a solution to the one-dimensional heat equation (5.25) with initial condition $u(0,x) = f(x)$ and Dirichlet boundary conditions $u(t,0) = g(t)$ and $u(t,L) = h(t)$. Let $v(t,x)$ denote another solution of the same initial-boundary value problem. Then consider

$$w(t,x) = u(t,x) - v(t,x).$$

It follows that $w(t,x)$ also satisfies the heat equation (5.25) but with zero initial condition and zero boundary conditions. By the Maximum Principle, $w(t,x) \le 0$ for $0 \le t \le T$ and $0 \le x \le L$, where $T > 0$. Similarly, by the Minimum Principle, $w(t,x) \ge 0$. Therefore, $w(t,x) = 0$, so that $u(t,x) = v(t,x)$ for all $t \ge 0$ and the initial-boundary value problem solution must be unique.

(b) *Heat kernel.* Consider now the one-dimensional heat equation (5.25) defined on the whole line, $-\infty < x < \infty$, with the initial condition $u(0,x) = f(x)$. The solution is given by

$$u(t,x) = \frac{1}{2\sqrt{\pi \alpha t}} \int_{-\infty}^{\infty} f(y) e^{\frac{-(x-y)^2}{4\alpha t}} \, dy.$$

This solution involves only the well-known "heat kernel" $K(t,z) = \frac{1}{2\sqrt{\pi \alpha t}} e^{\frac{-z^2}{4\alpha t}}$ and the initial data f.

A special case is when the initial function is the Dirac delta function, $f(y) = \delta(y)$ — the "delta" pulse of heat. Then the solution becomes

$$u(t,x) = \frac{1}{2\sqrt{\pi \alpha t}} e^{\frac{-x^2}{4\alpha t}}.$$

This solution exhibits what is referred to as an "infinite speed of propagation." We note that in actual fact heat does *not* propagate down a uniform rod with infinite velocity. While this solution *is* an exact solution to the one-dimensional heat equation, it is an inarguable manifestation of the fact that our carefully derived heat equation model is only an *approximation* to physical diffusion or transport of heat in a rod.

(c) Next consider the one-dimensional heat equation (5.25) defined on a semi-infinite domain $0 < x < \infty$ with the initial condition $u(0, x) = f(x)$. This is called the *semi-infinite slab* problem and is an idealization that is *sometimes* useful in practical applications. Exact solutions are known for the following special cases of boundary conditions:

(i) *Dirichlet* boundary condition case:

$$u(t, 0) = 0.$$

By defining an odd extension of the function $f(x)$ as

$$f_{\text{odd}}(x) = \begin{cases} f(x) & x > 0, \\ -f(x) & x < 0, \\ 0 & x = 0, \end{cases}$$

and using the result from part (b), one obtains an explicit formula for the solution $u(t, x)$ of the form

$$u(t, x) = \int_0^\infty (K(t, x - y) - K(t, x + y)) f(y) \, dy,$$

where $K(t, z)$ is the heat kernel given in part (b) above.

(ii) *Neumann* boundary condition case:

$$\frac{\partial u}{\partial x}(t, 0) = 0.$$

Since the derivative of an even function is odd, we use an even extension of $f(x)$. Using the same reasoning as in part (i), we find the solution to be

$$u(t, x) = \int_0^\infty (K(t, x - y) + K(t, x + y)) f(y) \, dy.$$

5.3 Experimental Modeling of Heat Transfer

In §5.2, the development of the mathematical model for heat conduction in a solid was described. As already mentioned in Chapter 1, one of the main difficulties in model development is the process of validating the mathematical model by comparing the model prediction to the field (or experimental) data. In addition, mathematical models contain parameters and coefficients that are not directly measurable in experiments (for example, the thermal diffusivity and the convective heat transfer coefficient). Hence, mathematical modelers and experimentalists must collaborate closely in order to develop effective quantitative models. That is, experiments must be carefully designed in order to provide sufficient data to accurately estimate model parameters and/or coefficients.

We now describe a physical experiment, which is relatively simple and is cost effective to set up, that can be used to validate the model development for the example of heat conduction in a rod previously derived. The general arrangement of equipment needed to set up this experiment is depicted in Figure 5.5. The actual heat experiment as set up in our own laboratory is shown in Figure 5.6. The experiment is carried out on a square metal bar of about 75cm length and 1 cm^2 cross section, with holes (in which thermocouples can be inserted) drilled about 4cm apart along the bar. In our lab, we use both copper and aluminum bars in the experiment to study the properties of different metals and how they affect the heat conduction. The heating element used is a soldering iron encapsulated cylindrical heater of 30W. There is a wide variety of temperature measuring devices available, such as thermocouples, solid state sensors, mercury in glass thermometers, and resistance temperature detectors. The device which is used in our experiment, and which is most suitable for this application, is the thermocouple. These thermocouples, which are mounted at multiple locations on the rod, are capable of measuring the temperature of both flat and curved metals, plastic or ceramic surfaces. To insure that thermal equilibrium between the rod and the thermocouple be established quickly (this also improves the accuracy of the measured temperature), the holes on the rod are drilled just large enough for the thermocouples to fit snugly. The temperature measurements are recorded in real time on a personal computer (PC) using a front-end analog multiplexer that quadruples the number of analog input signals. This arrangement allows up to 64 thermocouples to be used for simultaneous temperature measurements. In all of our experiments, 15 thermocouples were used. The analog signals are digitized by a MIO series multifunction data acquisition (DAQ) board. This board is plugged into one of the empty ISA slots of the PC.

5.3.1 The Thermocouple as a Temperature Measuring Device

The relationship between temperatures and thermocouple output voltages is highly nonlinear. The Seebeck coefficient, or voltage change per degree of temperature change, can vary by a factor of three or more over the operating temperature range of some thermocouples. For this reason, the temperature from thermocouple voltage readings must be approximated by polynomials.

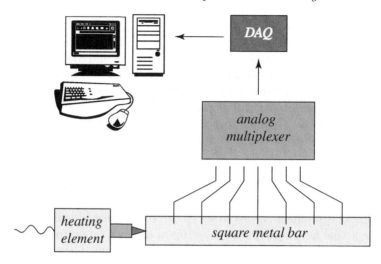

FIGURE 5.5:   Hardware connections used to validate the one-dimensional heat equation.

The polynomials, which can be used to convert the voltage readings in microvolts to degrees Celsius and vice versa, are given in Tables 2.2 and 2.3. These formulas are taken from [9]. The thermocouple used in this experiment is of type T, which has a fast response time, and is one of the oldest and most popular thermocouples for determining temperatures within the range from about 370 °C down to the triple point of neon (-248.5939 °C). Its positive thermoelement, TP, is typically copper of high electrical conductivity and low oxygen content (99.95% pure copper with an oxygen content varying from 0.02 to 0.07% — depending on sulfur content — and with other impurities totaling about 0.01%).

To compensate for the temperature difference between the measuring end and the cold junction (AMUX-64T multiplexer screw terminal), the following procedure can be used:

(i)   Translate the ambient temperature into the corresponding voltage using the polynomial in Table 5.3.

(ii)   Add the voltages from thermocouples readings to the voltage from step (i).

(iii)   Translate the voltage results from step (ii) into the temperatures using the polynomial from Table 5.2.

TABLE 5.2: Type T thermocouples: Coefficients of the approximate inverse function giving temperature u as a function of the thermoelectric voltage E in the specified temperature and and voltage ranges. The function is of the form: $u = c_0 + c_1 E + c_2 E^2 + \cdots + c_6 E^6$, where E is in microvolts and u is in degrees Celsius.

Temperature range:	0 °C to 400 °C
Voltage range:	$0\mu V$ to $20872\mu V$
$c_0 =$	0.000000
$c_1 =$	2.592800×10^{-2}
$c_2 =$	-7.602961×10^{-7}
$c_3 =$	4.637791×10^{-11}
$c_4 =$	$-2.165394 \times 10^{-15}$
$c_5 =$	6.048144×10^{-20}
$c_6 =$	$-7.293422 \times 10^{-25}$
Error range:	0.03 °C to -0.03°C

TABLE 5.3: Type T thermoucouples: Coefficients of the approximate function giving the thermoelectric voltage E as a function of temperature u in the specified temperature range. The function is of the form: $E = c_0 + c_1 u + c_2 u^2 + \cdots + c_8 u^8$, where E is in microvolts and u is in degrees Celsius.

Temperature range:	0 °C to 400 °C
$c_0 =$	0.000000
$c_1 =$	3.8748106364×10^1
$c_2 =$	$3.3292227880 \times 10^{-2}$
$c_3 =$	$2.0618243404 \times 10^{-4}$
$c_4 =$	$-2.1882256846 \times 10^{-6}$
$c_5 =$	$1.0996880928 \times 10^{-8}$
$c_6 =$	$-3.0815758772 \times 10^{-11}$
$c_7 =$	$4.5479135290 \times 10^{-14}$
$c_8 =$	$-2.7512901673 \times 10^{-17}$

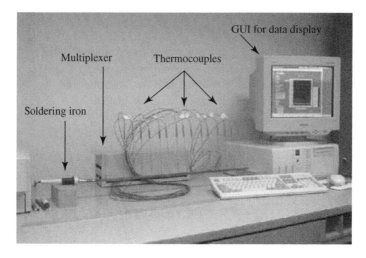

FIGURE 5.6:   Heat experiment as set up in our own laboratory.

TABLE 5.4:   Hardware equipment for thermal equipment.

Descriptive name	Probable brand (model)
• Super fast response time thermocouples	Omega (C01-K)
• PC data acquisition boards for ISA	National Instruments (AT-MIO-16E-10)
• Analog multiplexer with temperature sensor	National Instruments (AMUX-64T)
• PC computer	Pentium (or later processor)

5.3.2   Detailed Hardware and Software Lists

To carry out the experiments outlined here, a minimum of the following hardware and software is advisable (see Tables 5.4 and 5.5, respectively).

Given a lab set up of the type described above, students can carry out a project to determine thermal constants in the heat model. We suggest in some detail such a project.

Project: Thermal Experiment

The aim of this project is to validate the mathematical model for heat transfer in a rod using data collected from the experiment as described in §5.3. To this end, we recall from §5.2 and under the following assumptions

(i) Heat is transferred along the x-axis only,

TABLE 5.5: Software tools for thermal equipment.

Descriptive name	Brand
• Labsuite software for data acquisition and equipment control including LabVIEW for Windows	National Instruments
• MATLAB and Optimization toolbox for data and model analysis	The MathWorks, Inc.

(ii) Temperature is uniform over a cross-section and,

(iii) The rod is *perfectly insulated*,

a one-dimensional heat equation describing the heat conduction in a rod was developed. This has the form

$$\rho c_p \frac{\partial u(t,x)}{\partial t} = k \frac{\partial^2 u(t,x)}{\partial x^2},$$

where c_p is the specific heat, ρ is the mass density, and k is the thermal conductivity.

1. One possible set of boundary conditions for this experiment is:

$$k u_x(t,0) = Q, \qquad u(t,L) = u_{\text{ambient}},$$

where Q is an unknown constant determining the constant heat flux at the source, u_{ambient} is the known ambient (room) temperature, and L is the length of the rod. The specified boundary condition at $x = L$ implies that we have a constant temperature end. Is this an accurate assumption? If not, what would be a reasonable alternative?

Formulate the least squares problem to estimate the parameters k and Q using the *steady-state* temperature values along the rod. The optimization problem associated with the inverse least squares technique can be solved numerically by using MATLAB routines `fminsearch` or `fminu`. Note that because steady-state temperature values are used as data, we only have to solve a steady-state one dimensional heat equation, which is a linear two-point boundary value problem for which an exact solution can be derived. Derive this exact solution. Also, discuss all data or measurements that would be required in order to carry out the parameter estimation problem.

2. Collect steady-state values of the temperature distribution in a copper rod. Using this data set, estimate the unknown parameters k and Q. Note that both MATLAB routines `fminu` and `fminsearch` are iterative methods that require an initial guess for the parameter set. Are the

estimated parameters k and Q unique with respect to different initial guesses? Explain the results. Plot the solution of the mathematical model against the data. How well does the model fit the data?

3. To compare the rates of heat flow in different materials under the same conditions, set up a second experiment involving an aluminum bar. Repeat part 2. above. Are the estimated values of k and Q changed (or unchanged) for the two rods? How do thermal conductivities of copper and aluminum bar affect their heat conduction? Compare the values of k you obtain with those given in the literature for copper and aluminum.

4. Instead of assuming a nonhomogeneous Dirichlet boundary condition at $x = L$ as in part 1., consider a convective cooling boundary condition of the form

$$ku_x(t, L) = h[u_{\text{ambient}} - u(t, L)],$$

where h is the convective heat transfer coefficient. Repeat part 2. (where you now estimate k, h, and Q) and compare the results.

5. In practice, it is very difficult to ensure perfect insulation. In fact, in the design of the experiment, we have an uninsulated metal bar which is heated at one end and allows heat to escape along its entire length.

Under this new assumption of no insulation as well as assumptions (i) and (ii) above, derive the following equation for the conduction of heat in the rod:

$$\rho c_p \frac{\partial u(t, x)}{\partial t} = k\frac{\partial^2 u(t, x)}{\partial x^2} - \frac{2(a+b)}{ab}h(u(t, x) - u_{\text{ambient}}), \quad 0 \leq x \leq L,$$

where a and b are the dimensions of the cross-sectional area of the rod. The minus term on the right hand side of the above equation comes from the heat loss term along the length of the rod, which should be modeled by Newton's law of cooling.

Repeat steps 1.-4. using this model.

6. You should obtain improved fits to your data. Give a comparison and discussion of your findings for the two different models.

References

[1] H.T. Banks, S.R. Durso, M.A. Goodhart and K. Ito, Nonlinear exothermic contributions to radio-frequency bonding of adhesives, Center for Research in Scientific Computation, North Carolina State University, Technical Report, CRSC-TR98-24, 1998; *Nonlinear Analysis: Theory Method and Applications; Series B, 2*, 2001, pp. 257–286

[2] H.T. Banks and F. Kojima, Boundary shape identification problems in two-dimensional domains related to thermal testing of materials, *Quarterly of Applied Mathematics*, Vol. **XLVII**, No. 2, 1989, pp. 273–293.

[3] W. Boyce and R. Diprima, *Elementary Differential Equations and Boundary Value Problems*, John Wiley & Sons, Inc., Hoboken, 8th ed., 2004.

[4] R.B. Bird, W.E. Stewart and E.N. Lightfoot, *Transport Phenomena*, John Wiley & Sons, Inc., New York, 1960.

[5] C.J. Geankoplis, *Transport Processes and Unit Operations*, Prentice Hall, Englewood Cliffs, 1993.

[6] D.M. Heath, C.S. Welch and W.P. Winfree, Quantitative thermal diffusivity measurements of composites, in *Review of Progress in Quantitative Nondestructive Evaluation*, D.G. Thompson and D.E. Chimenti (eds.), Plenum Publ., **5B**, 1986, pp. 1125–1132.

[7] F. Incropera and D. Dewitt, *Fundamentals of Heat and Mass Transfer*, John Wiley & Sons, New York, 1990.

[8] W. Kaplan, *Advanced Calculus*, Addison-Wesley Publishing Co., Inc., New York, 1991.

[9] NIST Monograph 175, *Temperature-Electromotive Force Reference Functions and Tables for the Letter-Designated Thermocouple Types Based on the ITS-90*, NIST Monograph 175, 1993.

[10] W.A. Strauss, *Partial Differential Equations: An Introduction*, John Wiley & Sons, New York, 1992.

Chapter 6

Structural Modeling:
Force/Moments Balance

6.1 Motivation: Control of Acoustics/Structural Interactions

In many aircraft, a major component of the interior sound pressure field is due to structure-borne noise from the engines. For example, the rotating blades in turboprop engines generate low frequency, high displacement acoustic fields that couple (nonlinearly) with the fuselage dynamics. These mechanical vibrations produce, through interactions with the air inside the cabin, interior sound pressure oscillations. If these acoustic pressure fields are left uncontrolled, they can lead to undesirable conditions for the passengers.

One way to attenuate the interior acoustic pressure field is through the use of stiffer structures and acoustic damping material. However, this passive control technique increases the weight of the aircraft and reduces its fuel efficiency. Another approach is active or feedback control of noise in the interior acoustic cavity. There is a substantial literature on this subject including a frequency domain approach (see, e.g., [5, 9]) as well as a time domain approach (e.g., [1, 3]). For example, one strategy for active control of noise is through the use of secondary source techniques with the secondary noise based on feedback of noise levels in the acoustic cavity. In this approach, loudspeakers and microphones are strategically placed in the interior cavity where one can measure the pressure field. This information is used as feedback for the speakers which generate an interfering secondary field to reduce the total noise levels (primary and secondary sources) in certain critical zones. This strategy is local in nature and requires a large array of external control hardware.

Another strategy for reducing sound pressure levels is through the use of smart materials technology such as piezoceramic or electrostrictive elements. These piezoceramic patches when bonded to the fuselage act as an electro-mechanical transducer. That is, when excited by an electric field, the patch induces a strain in the material to which it is bonded and hence can be used as an actuator. On the other hand, if the bonded material undergoes a deformation, this produces a strain in the patch which results in a voltage across the patch (that is proportional to the the strain) and thereby permits the use of the patch as a mechanical sensor.

In [3] a model problem was considered that consists of an exterior noise source which is separated from an interior cavity by an active plate. As a 2-D analogue of the plate, the coupling boundary between the exterior noise source and the interior cavity is modeled by a fixed-end Euler-Bernoulli beam with Kelvin-Voigt damping (see Figure 6.1). The acoustic response inside the cavity is modeled by a linear wave equation.

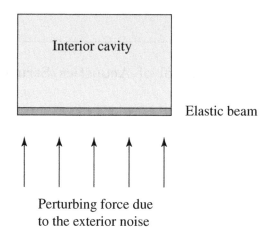

FIGURE 6.1: 2-D fluid/structure interaction system.

In this chapter we will consider the development of the Euler-Bernoulli model for the transverse displacement of a beam structure.

6.2 Introduction to Mechanics of Elastic Solids

In this section we briefly review the behavior of elastic bodies subjected to various types of loading. This field of study is known in the literature by several names including *strength of materials*, *mechanics of materials*, or *mechanics of deformable bodies*. Mechanics of solids is a fairly old and well established subject. It dates back to the work of Galileo in the early part of the seventeenth century. Of course much progress has been made since then, notably the development of the subject by French investigators such as Coulomb, Poisson, Navier, St. Venant, and Cauchy in the nineteenth century. Today there is substantial literature on this subject, see for example, [7, 10, 13,

14] and the references contained therein. A good understanding of mechanical behavior is essential for the development of the mathematical model describing the beam displacement to be discussed later in this chapter.

6.2.1 Normal Stress and Strain

The concepts of stress and strain can be illustrated using a *prismatic bar* (having uniform cross section throughout its length) with axial forces P applied at both ends. In Figure 6.2, the axial forces produce a uniform stretching of the bar, and the bar is said to be in tension.

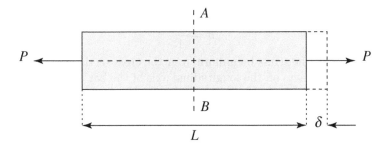

FIGURE 6.2: Prismatic bar deformation due to tensile forces.

To consider stresses and strains in this bar, we make an artificial plane cut (AB) that is perpendicular to the longitudinal axis of the bar (see Figure 6.2). This plane isolates the bar into two free bodies (parts of the bar to the left and to the right of the cut). Consider the part of the bar to the right of the cut as depicted in Figure 6.3. The tensile load P acts at the right end of this free body, while at the other end are forces representing the action of the removed part of the bar on the part that remains.

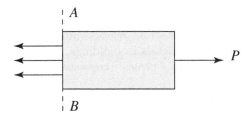

FIGURE 6.3: Normal stresses on the prismatic bar.

These forces are continuously distributed over the cross section. The intensity of this force, force per unit area, is called the *stress*, is denoted by σ, and is defined as

$$\sigma = \frac{\text{force}}{\text{area}} = \frac{F}{A}.$$

From the equilibrium of the body, the magnitude of F is equal to P and is in the opposite direction. This is indeed a consequence of Newton's Third Law which says to every action there is always an equal opposed reaction. Therefore,

$$\sigma = \frac{P}{A}.$$

When the bar is stretched or in tension, the stress is called *tensile stress*. On the other hand, if the forces are reversed in direction, causing the bar to be compressed, we obtain compressive stresses. A necessary condition for the formula

$$\sigma = \frac{P}{A}$$

to be valid is that the stress be uniform over the cross section of the bar. This is realized when P acts on the longitudinal axis of the bar as shown in Figure 6.2. If the load P is not on this longitudinal axis, bending of the bar will occur and the stress is more complicated to derive.

The elongation of the bar due to axial forces P is denoted by δ and the elongation per unit length

$$\varepsilon = \frac{\delta}{L}$$

is called the *strain*, which is a dimensionless quantity. Just as in the case of stress, there are *tensile* and *compressive strains*.

6.2.2   Stress and Strain Relationship (Hooke's Law)

The relationship between stress and strain in a particular material is determined by a *tensile test*. A material, usually a prismatic bar, is placed in a testing machine and subjected to a tension. The force on the bar and the elongation of the bar are measured as the load is increased. The stress in the bar is found by dividing the force by the cross-sectional area. The strain is found by dividing the elongation by the original length of the bar. By determining the stress and strain for various magnitudes of the load, we can plot a curve showing the stress-strain relationship. Structural steel, which is one of the most widely used metals in buildings, bridges, cranes, etc., when subjected to tensile forces has a typical stress-strain diagram as depicted in Figure 6.4.

In the diagram from the point 0 to A, the stress-strain relationship is linear (i.e., the stress and strain are proportional). Beyond A, the stress-strain relationship is nonlinear. The point A is called the *proportional limit*. Beyond A, increasing the loading increases the strain more rapidly than the stress

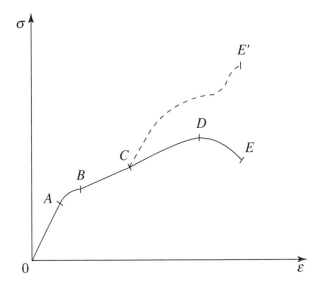

FIGURE 6.4:   Stress-strain diagram for a typical structural steel in tension.

until at point B when a considerable elongation begins to occur with a very small increase in the tensile force. This phenomenon is known as *yielding* of the material, and the stress at point B is known as the *yield point* (or *yield stress*). In the region BC, the material is said to become *plastic*. At point C, the material begins to *strain harden* which resists further loading increases. Thus, with further elongation, the stress increases and it reaches an *ultimate stress* at point D. Beyond D, further increase in the strain causes a reduction in the stress until the material breaks down at point E.

In the vicinity of the ultimate stress, a reduction in the cross-section area of the bar might occur (see Figure 6.5), called *necking*, which will have an effect on calculating the stress. In this case, the stress-strain curve follows the dashed line CE' as depicted in Figure 6.4.

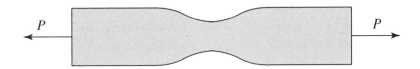

FIGURE 6.5:   Necking of a prismatic bar in tension.

We note that aluminum alloys have a more gradual transition from the linear region to the nonlinear region. In addition, both steel and aluminum undergo a large strain before failure and are thus classified as *ductile*. An advantage of ductility is that visible distortions may occur if the load becomes excessively large, thus providing means for a visual inspection to take remedial action before an actual fracture occurs. Finally, stress-strain diagrams for compression are different from the tension case. For example, ductile metal alloys such as steel, aluminum, and copper have proportional limits in compression very close to those in tension; however, when yielding begins, the behavior is quite different.

So far we have discussed the behavior of elastic solids subjected to tensile forces or compressive forces. Now let us consider what happens when the material is unloaded. In this case, the elongation will either partially or completely disappear. If the bar completely recovers its original shape when the load is removed, it is said to be *perfectly elastic*. Otherwise, it is known as *partially elastic*. In the latter case, the elongation that remains when the load is removed is called the *permanent set*.

For a perfectly elastic material, the process of loading and unloading can be repeated for successively higher values of the loading force. Eventuallly, a stress will be found for which a residual strain remains during unloading (i.e., the material becomes a partially elastic material). In this case, the stress which represents the upper limit of the elastic region is known as the *elastic limit*. For steel, as well as many other metals, the elastic limit and the proportional limit nearly coincide.

Many structural materials have an initial region on the stress-strain curve like the curve $0A$ in Figure 6.4. When a material behaves *elastically* and has a linear relationship between stress and strain, it is called a *linear elastic material*. Linear elasticity in the initial region of the stress-strain region is a property of many materials such as metals, plastics, wood, ceramics, etc.

The linear relationship between strain and stress can be described by

$$\sigma = E\varepsilon,$$

which is commonly known as *Hooke's Law*. Here, E is a constant of proportionality and is known as the *modulus of elasticity* or *Young's modulus*. It is given in units N/m^2 (the same as stress). By the definition of stress, we have

$$\sigma = \frac{P}{A} = E\frac{\delta}{L}.$$

Solving for δ we obtain

$$\delta = \frac{LP}{EA}.$$

Thus for a linear elastic bar, the elongation is directly proportional to L and P and inversely proportional to E and A. The product EA is known as the *axial rigidity* of the bar. In addition, the *flexibility* of the bar is defined as the deflection due to a unit value of the load and is given by $\frac{L}{EA}$. Similarly, the *stiffness* is defined to be the force required to produce a unit deflection and is given by $\frac{EA}{L}$, which is the reciprocal of the flexibility. These are important concepts in the analysis of structural materials.

When a prismatic bar is subjected to tensile loading, the axial elongation is accompanied by a lateral contraction, which is normal to the direction of the applied forces. Within the *elastic region*, the ratio

$$\nu = \frac{|\text{lateral strain}|}{|\text{axial strain}|}$$

is constant and is called *Poisson's ratio* named after the famous French mathematician S.P. Poisson (1781-1840). For *isotropic* materials (having the same elastic properties in all directions), he found $\nu = 0.25$.

6.2.3 Shear Stress and Strain

In the previous sections we described the behaviors of a prismatic bar under normal stresses (i.e., the loading forces act normal to the surface of the material). We now consider another kind of stress, called shear stress, that acts parallel to the surface of the material. To illustrate the concept of shear stress, we consider a bolted connection as depicted in Figure 6.6.

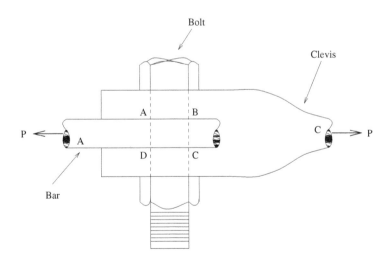

FIGURE 6.6: Bolt subjected to bearing stresses in a bolted connection.

Under the influence of the tensile forces P, the prismatic bar and the clevis will press against the bolt in shearing resulting in shearing stresses being developed against the bolt (see Figure 6.7). In addition, the bar and the clevis tend to shear the bolt across the sections AB and DC as depicted also in Figure 6.7. In this particular consideration, there are two planes of shear (along AB and DC), and so the bolt is said to be in double shear. Each of the shear forces S is indeed equal to $P/2$. The exact distribution of these shear stresses is not easily determined, but its average value is given by

$$\tau_{AVG} = \frac{S}{A}, \tag{6.1}$$

where A is the cross-sectional area of the bolt. Since shear stresses act tangential to the surface on which they act, they are also called *tangential stresses*. From equation (6.1), shear stresses, like normal stresses, represent intensity of force per unit area and thus they have the same unit $[force/area]$ as normal stresses.

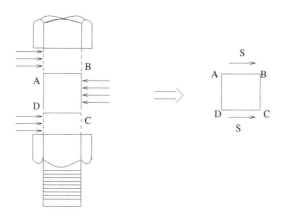

FIGURE 6.7:   Shearing stresses exerted on the bolt by the prismatic bar and the clevis.

To understand the deformation due to shear stresses, we next consider a cubical element with a shearing stress acting on the top plate as shown in Figure 6.8.

If there are no normal stresses, equal and opposite shear stresses must act on the bottom plate (otherwise, the block will move horizontally). These two shear stresses produce a moment which must be balanced by shearing stresses acting on the right and left plates (see Figure 6.9).

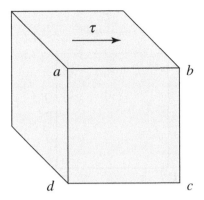

FIGURE 6.8: Shear stress acts on a rectangular cube.

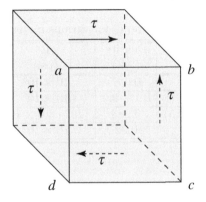

FIGURE 6.9: Shear stresses.

These shearing stresses will transform the square *abcd* into a rhombus as shown in Figure 6.10.

The angle γ, a measure of the distortion of the element due to shear, is called *shearing strain* and is given by

$$\tan \gamma = \frac{\beta}{|bc|}.$$

When the material behavior is in a linear elastic region, the stress-strain relationship, as discussed earlier, is linear as is the shear stress and shear strain relationship. In particular, we have Hooke's law for shear

$$\tau = G\gamma,$$

where G is the shear modulus of elasticity for the material. It is noted that E and G are not independent of each other. Indeed, for so-called Hookean

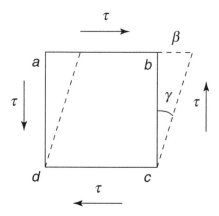

FIGURE 6.10:   Shear strains on the front side of the rhombus.

TABLE 6.1:   Values of E and G for various materials.

Material	Density $(\frac{lb}{in^3})$	E (psi)	G (psi)
Aluminum	0.097	10×10^6	4×10^6
Aluminum Alloys	$0.1 - 0.3$	10×10^6	4×10^6
Steel (mild high strength)	0.283	$(29 - 30) \times 10^6$	$(11 - 12) \times 10^6$
Copper	0.32	15×10^6	6×10^6

materials we have

$$G = \frac{E}{2(1 + \nu)}.$$

Values of E and G are listed in Table 6.1 for various commonly used materials [10].

6.3   Deformations of Beams

By the word *beam*, we mean a bar that is subjected to forces acting transversely to its axis. Beams are different from prismatic bars that are subjected to tensile and compressive forces because of the directions of the load that are applied to them. The loads on the bar act along its longitudinal axis. On the other hand, the loads on a beam act normal to its axis. Types of beams that are frequently studied include the cantilever beam and the simply supported (or simple) beam [10]. For a cantilever beam, one end of the beam is fixed and the other end is free (see Figure 6.11). At the fixed (or clamped) end, the beam can neither translate (horizontally or vertically) nor rotate. However, at the free end, it may do both.

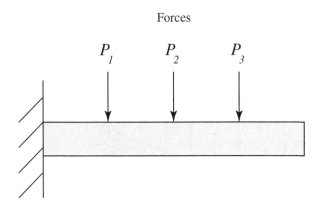

FIGURE 6.11: A cantilever beam.

Another example, shown in Figure 6.12, is a simply supported beam. Here, one end of the beam is supported by a pin and the other end by a roller. A pinned support is capable of resisting horizontal as well as vertical forces. Whereas, a roller support resists vertical forces only.

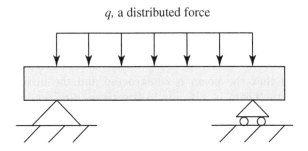

FIGURE 6.12: A simple beam.

For beams, loads can be *concentrated forces* such as P_1, P_2, and P_3 as depicted in Figure 6.11 or a *distributed force* such as q shown in Figure 6.12. Distributed loads are characterized by their intensity, which is the force per unit length (along the axis of the beam).

6.3.1   Differential Equations of Thin Beam Deflections

In this section we will develop general equations for the transverse deflection or displacement of the center line or axis of a thin cantilever beam. In particular, we consider a cantilever beam (depicted in Figure 6.13) with a tip mass at the free end that is subjected to a distributed force $f(t,x)$ along its axis, in units force per unit length (e.g., N/m).

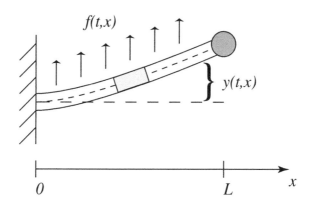

FIGURE 6.13:   A cantilever beam with a tip mass at the free end and is subjected to a distributed force f.

6.3.1.1   Force Balance

It is assumed that the beam is constructed and the in plane forces are applied so that the beam bends only in transverse direction (with no out of plane torsion or twisting about the axis of the beam). The beam is assumed *thin* and rigid so that its motion is completely characterized by the motion of its neutral axis, which in this case is the same as its centroid.

Let $y(t,x)$ denote the transverse displacement of the axis of the beam from its rest position. If we cut through the beam at ab (see Figure 6.14) and isolate the part of the beam to the right of the cut as a free body, we see that the action of the removed part (left side of the cut) upon the part of the beam to the right of the cut must be such as to hold the right side of the beam in force equilibrium. In Figure 6.14, M denotes the bending moments due to the applied load and S denotes the shear forces.

Consider an incremental element between x and $x + \Delta x$ of the beam $abdc$

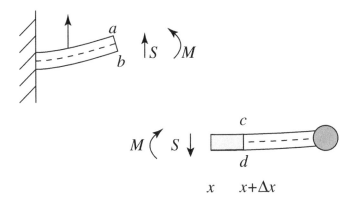

FIGURE 6.14: Shearing forces and moments on a cantilever beam with a tip mass at the free end.

as depicted in Figure 6.15. A general force balance yields

$$\int_{x}^{x+\Delta x} f(t,s)\,ds + S(t, x+\Delta x) - S(t,x) - f_I$$
$$- \int_{x}^{x+\Delta x} \gamma_d \frac{\partial y}{\partial t}(t,s)\,ds = 0, \quad (6.2)$$

where f_I represents the inertia force due to the mass of the differential element *abdc* and is given by a sum of point inertia forces f_i

$$f_I = \int_{x}^{x+\Delta x} f_i(t,s)\,ds = \int_{x}^{x+\Delta x} \rho \frac{\partial^2 y}{\partial t^2}(t,s)\,ds.$$

Here, ρ denotes the linear mass density in units mass per unit length. The term

$$f_D = \int_{x}^{x+\Delta x} f_d(t,s)\,ds = \int_{x}^{x+\Delta x} \gamma_d \frac{\partial y}{\partial t}(t,s)\,ds$$

in equation (6.2) represents the sum of viscous air damping forces $f_d(t,s) = \gamma_d \frac{\partial y}{\partial t}(t,s)$. After rearranging terms and dividing both sides of equation (6.2) by Δx, we obtain

$$\frac{1}{\Delta x} \int_{x}^{x+\Delta x} f(t,s)\,ds + \frac{S(t, x+\Delta x) - S(t,s)}{\Delta x}$$

$$- \frac{1}{\Delta x} \int_{x}^{x+\Delta x} \rho \frac{\partial^2 y}{\partial t^2}(t,s)\,ds - \frac{1}{\Delta x} \int_{x}^{x+\Delta x} \gamma_d \frac{\partial y}{\partial t}(t,s)\,ds = 0.$$

Now taking the limit as $\Delta x \to 0$ we find

$$f(t, x) + \frac{\partial}{\partial x} S(t, x) - \rho \frac{\partial^2 y}{\partial t^2} - \gamma_d \frac{\partial y}{\partial t} = 0. \tag{6.3}$$

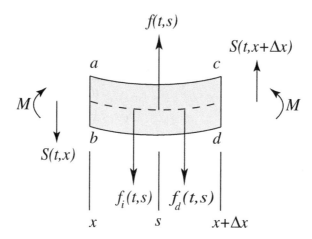

FIGURE 6.15: Force balance on an incremental element of the beam.

Equation (6.3) describes the transverse beam displacement y in terms of the loading force f and the shear stress S. Next we will derive an expression for the shear stress S by satisfying another requirement for equilibrium conditions for the beam segment *abdc*, namely, moment balance.

6.3.1.2 Moment Balance

In the previous section force balance, $\sum f_k = 0$, assures that one of two requirements for equilibrium of a beam segment is met. In particular, the equation $\sum f_k = 0$ is satisfied by relying on the existence of shear forces at a section of a beam. The remaining condition of static equilibrium is $\sum M_k = 0$. This, in general, can be satisfied by considering the internal resisting moment within the cross-sectional area of the cut to balance the external moment caused by the load (see Figure 6.15). The internal moment must act in a direction opposite to the external moments to satisfy the moment balance equation ($\sum M_k = 0$). These moments bend the beam in the plane of the loads and are generally referred to as *bending moments* about points on the neutral axis (or more precisely, about axes perpendicular to the plane of loading passing through points on the neutral axis).

To maintain the equilibrium of the segment *abdc*, we consider moment balance about the plane *ac* (actually about the neutral axis point midway between *a* and *c*). Adapting the convention that positive (+) moment is counterclock-wise, we obtain

$$-M(t, x) + M(t, x + \Delta x) + S(t, x + \Delta x)\Delta x$$

$$+ \int_x^{x+\Delta x} [f(t, s) - f_i(t, s) - f_d(t, s)](s - x)ds = 0. \quad (6.4)$$

After rearranging terms and dividing both sides of equation (6.4) by Δx, we arrive at (taking $F \equiv f - f_i - f_d$)

$$\frac{M(t, x + \Delta x) - M(t, x)}{\Delta x} + S(t, x + \Delta x) + \frac{1}{\Delta x} \int_x^{x+\Delta x} F(t, s)(s - x)ds = 0.$$

In the limit, as $\Delta x \to 0$, we have:

$$\frac{\partial}{\partial x} M(t, x) + S(t, x) = 0,$$

or

$$S(t, x) = -\frac{\partial}{\partial x} M(t, x).$$

Substituting the above expression for shear force into equation (6.3), we obtain the expression describing the dynamics of the beam displacement

$$\rho \frac{\partial^2 y}{\partial t^2} + \gamma_d \frac{\partial y}{\partial t} + \frac{\partial^2}{\partial x^2} M(t, x) = f(t, x). \quad (6.5)$$

For the above equation to be useful, we need to derive an expression for the bending moment M in terms of the beam displacement y.

6.3.1.3 Moment Computation

To continue theoretical development of in-plane beam *deflections*, the geom-etry of beam local *deformation* will be considered in this section. That is, to describe the center line displacement, we need to understand internal moments which are produced by local deformations (elongations and compressions) in the plane of displacement motion. We begin with the following fundamental assumptions (which are often called the *Euler-Bernoulli assumptions*):

(i) Plane sections remain planes during deformation;

(ii) Transverse displacement is *small* compared to the length of the beam;

(iii) Shear *deformation* is generally very small and will be neglected (no internal shear).

We note that for a beam where $L = 10h$, that is, a slender beam, the deflection due to shear is less than 1% [10].

In Figure 6.16 an initially straight segment of the beam (top figure) is shown after it undergoes deformation (bottom edge) and compression (top edge). The deflected neutral axis (that portion of the beam that does not undergo compression or elongation) of the beam OO' is bent into a curve with radius R. The center of curvature can be found by extending any two adjoining sections such as $a'e$ and $b'f$.

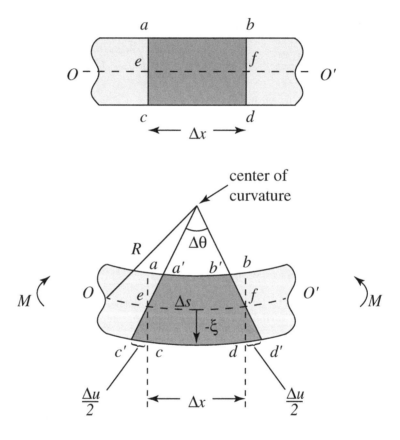

FIGURE 6.16: Local deformation of a segment of the beam due to bending.

The arc length of ef is given by

$$\Delta s = R\Delta\theta,$$

where $\Delta\theta$ is the angle between two adjoining sections $a'e$ and $b'f$. From the above formula we obtain

$$\frac{\Delta\theta}{\Delta s} = \frac{1}{R} = \kappa,$$

where κ denotes the *curvature* of the center line or neutral axis of the beam. In addition, the strained fiber at distance $-\xi$ from the neutral axis has elongation (half length)

$$\frac{\Delta u}{2} = -\xi \frac{\Delta \theta}{2}. \tag{6.6}$$

We note that in the above formula the negative sign $(-)$ of ξ is consistent with the elongation due to the deformation as depicted in Figure 6.16. Since the fiber ef is not strained, we obtain

$$\Delta s = \Delta x.$$

Substituting this expression into equation (6.6), we obtain

$$\frac{\Delta u}{\Delta x} = -\xi \frac{\Delta \theta}{\Delta s} = -\xi \kappa.$$

The above formula can then be used to compute the strain ε to be

$$\varepsilon = \frac{\Delta u}{\Delta x} = -\xi \frac{\Delta \theta}{\Delta x} = -\xi \kappa, \tag{6.7}$$

where Δu is the elongation due to deformation and Δx is the original length of the beam segment.

Next, by Hooke's law we obtain

$$\sigma = E\varepsilon$$
$$= -E\xi\kappa$$
$$= -\beta\xi, \tag{6.8}$$

where $\beta = E\kappa$. Equations (6.7) and (6.8) show that normal strain and stress vary linearly with their respective distances ξ from the neutral axis (see also Figure 6.17). The absolute maximum stress and strain occur at the edges of the beam.

Balancing the moments about the point e at x, we obtain

$$M + \int_A \sigma \xi \, dA = 0,$$

where σdA is the infinitesimal force acting on a cross-sectional element of area dA. Substituting expression (6.8) for the stress σ we find

$$M = \int_A \beta \xi^2 \, dA$$
$$= \beta \int_A \xi^2 \, dA, \tag{6.9}$$

where $\int_A \xi^2 \, dA$ is the moment of inertia I of the cross-sectional area about the centroidal axis and ξ is measured from this axis. In fact, the neutral axis

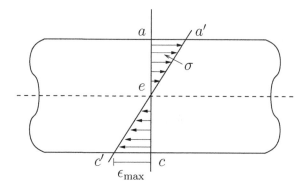

FIGURE 6.17: Stress and strain as functions of distances from the neutral axis at the point x (or e) on the neutral axis.

of the beam passes through the centroid of the cross-sectional area. This can be seen from the following observation: by force balancing, we have

$$\sum F_k = 0,$$

or

$$\int_A \sigma \, dA = 0.$$

Using equation (6.8) we obtain

$$\beta \int \xi \, dA = 0,$$

which implies

$$\int \xi \, dA = 0.$$

Therefore, we have

$$\int \xi \, dA = \bar{\xi} A = 0,$$

or

$$\bar{\xi} = 0.$$

That is, the distance from the neutral axis to the centroid $\bar{\xi}$ is zero, or the neutral axis passes through the centroid of the cross-sectional area.

From equation (6.9) and the definition of the parameter β in equation (6.8), we obtain the expression for the bending moment

$$M = EI\kappa. \tag{6.10}$$

Next, we recall from elementary calculus that for a curve (in our case the displacement of the center line of the beam) given by $z = z(x)$, the curvature is given by the formula

$$\kappa = \frac{\frac{d^2 z}{dx^2}}{\left[1 + \left(\frac{dz}{dx}\right)^2\right]^{\frac{3}{2}}}.$$

If $\frac{dz}{dx} \ll 1$ (a condition justified by the small displacement assumption (ii) at the beginning of this section), we obtain

$$\kappa \approx \frac{d^2 z}{dx^2}. \tag{6.11}$$

Hence the bending moment is expressed in term of the beam displacement $y(t, x)$ by the equation

$$M_{\text{int}} = M_{\text{bending}} = EI \frac{\partial^2 y}{\partial x^2}. \tag{6.12}$$

Before closing this section, we note that for a beam with a rectangular cross-sectional area (thickness h and width b) as depicted in Figure 6.18, the moment of inertia I can be computed as a function of the cross-sectional width and height as

$$I = \int_A \xi^2 dA$$

$$= \int_{-\frac{h}{2}}^{\frac{h}{2}} \int_{-\frac{b}{2}}^{\frac{b}{2}} \xi^2 dz d\xi$$

$$= b \int_{-\frac{h}{2}}^{\frac{h}{2}} \xi^2 d\xi$$

$$= b \frac{h^3}{12}.$$

Finally, the formula (6.12) that we derived above did not take into account any resistance of material (internal resistance) to bending. For a material known as a viscous material or viscoelastic material, the stress depends on strain rate as well as strain [10]. That is,

$$\sigma = E\varepsilon + c_D \dot{\varepsilon}.$$

From moment balance we have for such materials

$$M = -\int_A \sigma \xi \, dA$$

$$= -\int_A (E\varepsilon + c_D \dot{\varepsilon}) \xi \, dA.$$

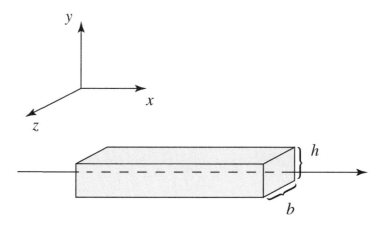

FIGURE 6.18:   Segment of a beam with a rectangular cross-sectional area.

Substituting equations (6.7) and (6.11), after some simple calculations we obtain

$$M = M_{\text{bending}} + M_{\text{damping}}$$
$$\approx EI\frac{\partial^2 y}{\partial x^2} + c_D I\frac{\partial^3 y}{\partial x^2 \partial t}. \tag{6.13}$$

It is seen from the above equation that for a viscoelastic material, the moment is the sum of two terms: the first term is the standard internal (bending) moment and the second term is due to the *structural damping* (which is also referred to as *Kelvin-Voigt* damping). There are other types of damping models that one encounters frequently in practice. For example, for composite fiberous materials

$$M_{\text{damping}}(t, x) = \int_x^l d\xi \int_0^l b(\xi, \theta) \left[\frac{\partial^2 y}{\partial \xi \partial t}(t, \xi) - \frac{\partial^2 y}{\partial \theta \partial t}(t, \theta)\right] d\theta.$$

This is known as *spatial hysteresis damping*. It follows from the above formula that the shear force is given by

$$S(t, x) = -\frac{\partial M}{\partial x}(t, x) = \int_0^l b(x, \theta) \left[\frac{\partial^2 y}{\partial x \partial t}(t, x) - \frac{\partial^2 y}{\partial \theta \partial t}(t, \theta)\right] d\theta.$$

Another type of damping known as *time hysteresis damping* is used to model many viscoelastic materials such as rubbers and other polymers, electromagnetic materials, and "smart materials" such as ferroelectrics and ferromagnetics. These materials exhibit hysteresis in their stress-strain curves. In particular, the stress-strain relationship has the form

$$\sigma = E\varepsilon + E_1\varepsilon_1,$$

where ε_1 is called the "internal" strain variable. One simple model for the internal strain variable is

$$\dot{\varepsilon}_1 + c\varepsilon_1 = k\varepsilon; \qquad \varepsilon_1(0) = 0.$$

Expressing the solution to the above first order linear differential equation by the variation of constants formula and substituting it into the stress-strain relationship, we obtain

$$\sigma(t) = E\varepsilon + E_1 \int_0^t e^{-c(t-s)} k\varepsilon(s)\, ds$$

$$= E\varepsilon(t) + \int_0^t K(t-s)\varepsilon(s)\, ds, \tag{6.14}$$

where $K(\cdot)$, the kernel function, is given by the exponential function for the simple internal strain model given above. Using the expression

$$M = -\int_A \sigma\xi\, dA$$

and equations (6.7) and (6.11), we obtain the following formula for the moment

$$M(t,x) = EI\frac{\partial^2 y}{\partial x^2}(t,x) + \int_0^t \hat{K}(t-s)\frac{\partial^2 y}{\partial x^2}(s,x)\, ds$$

$$= M_{\text{bending}} + M_{\text{hysteresis damping}}$$

6.3.1.4 Initial Conditions

If one does not take into account any resistance of material to bending, the moment is given by

$$M = M_{\text{int}} = M_{\text{bending}} = EI\frac{\partial^2 y}{\partial x^2}.$$

Substituting this expression into the dynamical equation (6.5) for transverse beam displacement, we have

$$\rho\frac{\partial^2 y}{\partial t^2} + \gamma_d\frac{\partial y}{\partial t} + \frac{\partial^2}{\partial x^2}\left(EI\frac{\partial^2 y}{\partial x^2}\right) = f(t,x). \tag{6.15}$$

This dynamical equation is second order in time and fourth order in space. Therefore, for well posedness of solutions, two initial conditions are required. These are usually initial displacement and initial velocity which are given by

$$y(0,x) = y_1(x),$$

$$\frac{\partial}{\partial t}y(0,x) = y_2(x),$$

where y_1 and y_2 are known functions.

6.3.1.5 Boundary Conditions

From the discussions above, we found that the dynamical equation (6.15) for the cantilever beam displacement is fourth order in space. Therefore, four boundary conditions need to be specified. Two of these boundary conditions are naturally specified at the fixed (clamped) end. This type of support resists a displacement in any direction as well as a slope in the beam. Therefore, we have the following mathematical conditions:

- *Fixed end support* $(x = 0)$:

$$y(t, 0) = 0$$
$$\frac{\partial}{\partial x} y(t, 0) = 0.$$

There are other types of supports used frequently for beams loaded with forces acting in the same plane. These include:

- *Pinned end support* $(x = 0)$:
  Since a pinned (or simple) end support (shown in Figure 6.19) resists a force acting in any direction of the plane, both the displacement and moment are specified to be zero. That is,

$$y(t, 0) = 0 = M(t, 0).$$

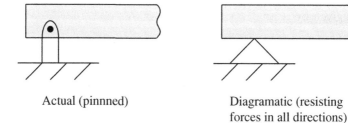

Actual (pinnned)   Diagramatic (resisting forces in all directions)

FIGURE 6.19: Pinned end support.

- *Free end support* $(x = 0)$:
  For this type of support, the moment as well as the shear force are zero. Thus,

$$M(t, 0) = \frac{\partial M}{\partial x}(t, 0) = 0.$$

- *Frictionless ring or roller end support* $(x = 0)$:
  This type of support (see Figure 6.20) allows vertical forces, resisting longitudinal forces only. Therefore, both the slope and the shear force are zero as specified by

$$\frac{\partial y}{\partial x}(t, 0) = \frac{\partial M}{\partial x}(t, 0) = 0.$$

(resisting force in only one direction --
the longitudinal direction)

FIGURE 6.20: Frictionless roller end support.

- *Tip mass* $(x = L)$:
  To allow for a more general (and often useful) formulation, we consider next a cantilevered beam with a tip mass at the free end as depicted in Figure 6.21. Figure 6.22 shows a segment of the beam at the tip end

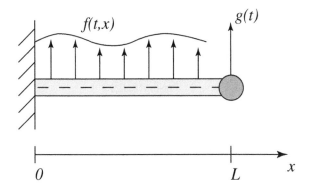

FIGURE 6.21: Cantilever beam with a tip mass.

after it undergoes deformation due to rotational inertia.

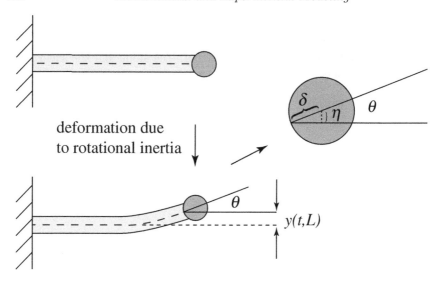

FIGURE 6.22: Local deformation of the cantilever beam with tip mass.

Since θ in Figure 6.22 is small by the Euler-Bernoulli assumption (ii), we have

$$\begin{aligned}
\tan\theta &= \frac{\sin\theta}{\cos\theta}\\
&\approx \sin\theta\\
&= \frac{\eta}{\delta},
\end{aligned}$$

which implies that

$$\eta = \delta\tan\theta.$$

Hence, the total transverse displacement of the center of mass of the tip body is

$$\eta + y(t, L).$$

Considering the dynamics at the tip mass and applying force balance including the inertia force at the tip (see Figure 6.23), we find

$$-S(t, L) + g(t) - (f_{tip})_{\mathrm{I}} = 0,$$

or,

$$-S(t, L) + g(t) - m\frac{\partial^2}{\partial t^2}[\eta + y(t, L)] = 0,$$

where m is the tip body mass and g is the resultant external force applied to the center of mass of the tip body. Thus, after arranging terms, we

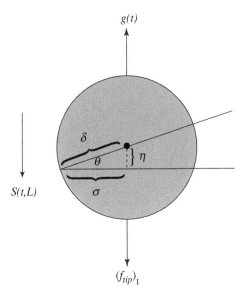

FIGURE 6.23: Force balance at the tip mass.

obtain

$$-\frac{\partial}{\partial x}M(t,L) + m\frac{\partial^2}{\partial t^2}[\delta\tan\theta + y(t,L)] = g(t).$$

It remains to determine $\tan\theta$. Consider a differential beam element *abdc* as before. The deformation of the beam due to the rotation of the beam cross section is illustrated in Figure 6.24. From Figure 6.24, it follows that

$$\theta \approx \tan\theta$$
$$\approx \frac{y(t, x + \Delta x) - y(t, x)}{\Delta x}$$
$$\approx \frac{\partial y}{\partial x},$$

where Δx is small. Here we utilize the assumptions that plane sections remain planes and that there is no shear deformation. Hence the boundary condition at $x = L$ is given by

$$\frac{\partial M}{\partial x}(t,L) + m\frac{\partial^2}{\partial t^2}\left[\delta\frac{\partial}{\partial x}y(t,L) + y(t,L)\right] = g(t).$$

It should be emphasized that if we also consider beam deformation due to shearing, the deformation dynamics can also be written but are much more complicated.

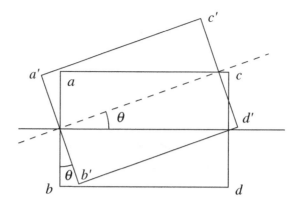

FIGURE 6.24: Deformation of the beam due to the rotation of the beam cross section.

To obtain the second boundary condition at $x = L$, we will now apply Newton's second rotational law (the sum of all moments equals to the product of the moment of inertia and the angular acceleration). Note that since it is a free end at the tip mass, it is free to rotate. Hence we have

$$\sum M_I = J\alpha(t). \qquad (6.16)$$

This expression is the counterpart to Newton's second law for rectilinear motion, which states that the sum of all applied forces equals to the product of mass and acceleration. In equation (6.16), J is the moment of inertia about the axis of rotation (which describes the spread of distribution of a region about an axis) and is given by

$$J = \int\int_V \int \bar{\rho}r^2 dV,$$

where $\bar{\rho}$ is the tip mass density. In addition, the angular acceleration, $\alpha(t)$, can be computed from the following expressions

$$
\begin{aligned}
\alpha(t) &= \frac{d}{dt}(\frac{d\theta}{dt}), \\
&\approx \frac{d}{dt}(\frac{d}{dt}\frac{\partial y}{\partial x}(t, L)), \\
&= \frac{\partial^3}{\partial t^2 \partial x}y(t, L).
\end{aligned}
\qquad (6.17)
$$

Finally, from Figure 6.25, we obtain that $\sum M_I = -M(t, L) + S(t, L)\delta + h(t)$, where h is the external applied moment about the center of mass of the tip body. Substituting this expression for the moment of inertia as well as the formula for the angular acceleration (6.17) into Newton's second rotational law (6.16), we obtain

$$J\frac{\partial^3}{\partial t^2 \partial x}y(t, L) = -M(t, L) + S(t, L)\delta + h(t)$$

or

$$J\frac{\partial^3}{\partial t^2 \partial x}y(t, L) + \delta\frac{\partial}{\partial x}M(t, L) + M(t, L) = h(t),$$

which gives us the required second boundary condition at $x = L$.

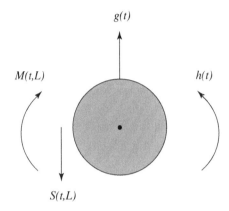

FIGURE 6.25: Moment balance at the tip mass.

6.4 Separation of Variables: Modes and Mode Shapes

During the last two centuries several methods have been developed for solving the types of partial differential equations such as those describing transverse beam displacement (e.g., (6.15)). Among these, in this section we will consider the method of separation of variables. This is perhaps the oldest systematic and relatively simple method for solving partial differential equations. It has been used since 1750 by D'Alembert, D. Bernoulli, and Euler in their studies of the wave equation (to be discussed later in this manuscript). In essence, the principal feature of this method is the replacement of the partial differential equation by a family of ordinary differential equations. The solution of the partial differential equation is then expressed, in general, as an infinite sum of solutions to these ordinary differential equations, usually in terms of trigonometric functions. Because of its relative simplicity and many important and practical problems to which it is applicable, the method of separation of variables has became one of the classical techniques of mathematical physics.

We begin by considering a simply supported (pinned-pinned) and undamped beam with no forcing function. Its transverse displacement satisfies

$$\rho \ddot{y} + (EIy'')'' = 0 \tag{6.18}$$

with boundary conditions

$$y(t,0) = y(t,L) = 0, \tag{6.19}$$

$$M(t,0) = M(t,L) = 0, \tag{6.20}$$

where $M(t,x) = EI\frac{\partial^2 y}{\partial x^2}$ and L is the length of the beam. Here $\dot{y}$ denotes the temporal derivative $\frac{\partial y}{\partial t}$, and $y' = \frac{\partial y}{\partial x}$, the spatial derivative. The initial conditions are given by

$$y(0,x) = y_1(x), \tag{6.21}$$

$$\dot{y}(0,x) = y_2(x), 0 \le x \le L. \tag{6.22}$$

The simply supported beam problem (6.18)-(6.22) is a linear homogeneous differential equation with linear boundary conditions. This suggests that one might want to seek solutions of the differential equation and boundary conditions, and then superpose them to satisfy the initial condition. In particular, we will apply the method of separation of variables to find these solutions.

In the method of separation of variables, one seeks a solution to the partial differential equation of the form

$$y(t,x) = w(t)\phi(x). \tag{6.23}$$

Substituting equation (6.23) for y into the differential equation (6.18), we obtain

$$\rho \ddot{w}\phi + w(EI\phi'')'' = 0,$$

or, equivalently,

$$-\frac{\ddot{w}}{w} = \frac{(EI\phi'')''}{\rho\phi}. \tag{6.24}$$

The variables in equation (6.24) are now separated if ρ and EI depend only on x and not on t; that is, the left term depends only on t and the right term depends only on x. Since the equation (6.24) is valid for $0 < x < L$ and $t > 0$, it is necessary that both sides of equation (6.24) be equal to the same constant, which we denote by β. Thus equation (6.24) becomes

$$-\frac{\ddot{w}}{w} = \frac{(EI\phi'')''}{\rho\phi} = \beta$$

from which we obtain the following two ordinary differential equations for $w(t)$ and $\phi(x)$:

$$\ddot{w} + \beta w = 0, \tag{6.25}$$

$$\frac{(EI\phi'')''}{\rho} - \beta\phi = 0. \tag{6.26}$$

Consequently, the partial differential equation (6.18) is replaced by two ordinary differential equations. If we assume that EI is constant, each of these equations can be readily solved for any constant β. However, we are only interested in those solutions of equation (6.18) that also satisfy the boundary conditions (6.19) and (6.20). This, in turn, restricts the possible values of β as we shall see below.

Substituting for $y(t, x)$ in equation (6.23) the boundary condition at $x = 0$, we obtain

$$y(t, 0) = w(t)\phi(0) = 0, \quad EI\frac{\partial^2 y}{\partial x^2}(t, 0) = EIw(t)\phi''(0) = 0. \tag{6.27}$$

Expressions given by equation (6.27) are satisfied if $w(t) = 0$ for all t or $\phi(0) = \phi''(0) = 0$. However, the condition $w(t) = 0$ for all t would imply that $y(t, x) = 0$. This is not acceptable since it does not satisfy the initial conditions (6.21)-(6.22) except in the trivial case $y_1 = y_2 = 0$. Therefore, equations (6.27) are satisfied by requiring that

$$\phi(0) = \phi''(0) = 0. \tag{6.28}$$

Similarly, substituting for $y(t, x)$ in equation (6.23) the boundary condition at $x = L$ we find

$$\phi(L) = \phi''(L) = 0. \tag{6.29}$$

Returning to (6.26), we have

$$\phi'''' = \frac{\beta\rho}{EI}\phi. \tag{6.30}$$

Now, multiplying both sides of equation (6.30) by ϕ, integrating both sides of the resulting equation from 0 to L, and then integrating by parts twice from the left side we obtain

$$\phi'''(L)\phi(L) - \phi'''(0)\phi(0) - \phi''(L)\phi'(L)$$
$$+ \phi''(0)\phi'(0) + \int_0^L (\phi''(x))^2 \, dx = \int_0^L \frac{\beta\rho}{EI}\phi^2(x) \, dx. \tag{6.31}$$

Substituting the boundary conditions (6.28)-(6.29) into equation (6.31) we obtain

$$\int_0^L (\phi''(x))^2 \, dx = \int_0^L \frac{\beta\rho}{EI}\phi^2(x) \, dx. \tag{6.32}$$

Since EI and ρ are all positive physical quantities, we must have $\beta = 0$ or $\beta > 0$. We note that if $\beta = 0$, then the solution to the equation (6.31) with boundary conditions (6.28)-(6.29) is $\phi(x) = 0$ for all $0 \le x \le L$. This would also imply that the solution $y(t, x) = 0$, which is unacceptable since this solution does not satisfy the initial conditions (6.21)-(6.22) except in the trivial case.

Summarizing our results to this point, we have shown that we can satisfy the boundary condition (6.20) for nontrivial solutions only if the separation constant β is positive. The corresponding solution to the linear, constant coefficient equation (6.30) can be now readily obtained from standard methods of ordinary differential equations and is given by

$$\phi(x) = a\cos(\xi x) + b\cosh(\xi x) + c\sin(\xi x) + d\sinh(\xi x), \qquad (6.33)$$

where $\xi = \left(\frac{\beta\rho}{EI}\right)^{1/4}$. Next, applying the boundary conditions (6.28) at $x = 0$, we have

$$a = b = 0.$$

In addition, applying the boundary conditions (6.29) at $x = L$, we find

$$d = 0, \quad c\sin(\xi L) = 0.$$

Hence, either $c = 0$ or $\sin(\xi L) = 0$. But $c = 0$ implies that $\phi(x) = 0$ and consequently $y(t, x) = 0$, which is again unacceptable for nontrivial solutions. Hence, we must have $\sin(\xi L) = 0$ and from this condition, we obtain

$$\xi L = n\pi, \quad n = 1, 2, \dots. \qquad (6.34)$$

We have thus shown that we can satisfy the boundary condition (6.20) only if the separation constant β is positive and is given by

$$\beta_n = \frac{EI}{\rho}\left(\frac{n\pi}{L}\right)^4, \quad n = 1, 2, \dots, \qquad (6.35)$$

where we used $\xi = \left(\frac{\beta\rho}{EI}\right)^{1/4}$. In addition, the functions

$$\phi_n(x) = \sin\left(\frac{n\pi}{L}x\right), \quad n = 1, 2, \dots, \qquad (6.36)$$

satisfy the boundary conditions (6.28)-(6.29) and the differential equation (6.30). These functions are called *eigenfunctions* as well as *mode shapes* in engineering literature. Next, upon substituting the values of β given by equation (6.35) into the differential equation (6.25), we find that $w(t)$ is given by

$$w_n(t) = A_n\cos\omega_n t + B_n\sin\omega_n t, \qquad (6.37)$$

where $\omega_n = \sqrt{\beta_n} = \sqrt{\frac{EI}{\rho} \frac{n^2 \pi^2}{L^2}}$, and A_n and B_n are arbitrary constants for $n = 1, 2, \ldots$. The constants ω_n are called *modes* or *natural frequencies*. Therefore, we conclude that the functions

$$y_n(t, x) = (A_n \cos \omega_n t + B_n \sin \omega_n t) \sin \left(\frac{n\pi}{L} x\right), \quad n = 1, 2, \ldots, \quad (6.38)$$

satisfy the differential equation (6.18) and the boundary conditions (6.20). However, since the differential equation and the boundary conditions are linear and homogeneous, by the superposition principle, any linear combination of the $y_n(t, x)$ also satisfies the differential equation and boundary conditions. Hence, we have

$$y(t, x) = \sum_{n=1}^{\infty} \hat{c}_n y_n(t, x)$$

$$= \sum_{n=1}^{\infty} (A_n \cos \omega_n t + B_n \sin \omega_n t)) \sin \left(\frac{n\pi}{L} x\right), \quad (6.39)$$

where, for simplicity, we absorbed the constants of proportionality $\hat{c}_n$ into A_n and B_n. It now only remains to satisfy the initial conditions (6.21)-(6.22).

That is, upon substituting the initial conditions into the equation (6.39) for $y(t, x)$, we must have

$$y_1(x) = \sum_{n=1}^{\infty} A_n \sin \left(\frac{n\pi}{L} x\right) \quad (6.40)$$

and

$$y_2(x) = \sum_{n=1}^{\infty} \omega_n B_n \sin \left(\frac{n\pi}{L} x\right). \quad (6.41)$$

These infinite series for y_1 and y_2 are the well known Fourier series (see Appendix A). The coefficients A_n and B_n are readily obtained from the following Euler-Fourier formulas [4]:

$$A_n = \frac{2}{L} \int_0^L y_1(x) \sin \left(\frac{n\pi}{L} x\right) dx,$$

$$B_n = \frac{2}{\omega_n L} \int_0^L y_2(x) \sin \left(\frac{n\pi}{L} x\right) dx. \quad (6.42)$$

These formulas as well as conditions on the functions y_1, y_2 that guarantee convergence of a Fourier series to the function from which its coefficients were computed can be found in most books on advanced calculus or applied mathematics (e.g., [6, 8]) (a brief introduction to the Fourier series and Fourier transform are given in Appendix A).

The method of separation of variables can also be used to solve the damped beam vibration problems with other boundary conditions than the pinned-pinned type given by equations (6.19)-(6.20). For example, in the case of internal damping such as Kelvin-Voigt damping (6.13), the equation for the transverse beam displacement is given by

$$\rho \ddot{y} + (EIy'')'' + (c_D I \dot{y}'')'' = 0. \tag{6.43}$$

Without loss of generality, we assume that boundary conditions and initial conditions are the same as in the undamped case and are given by equations (6.19)-(6.20) and (6.21)-(6.22), respectively (the project assignment at the end of this chapter deals with the boundary condition case associated with the cantilever beam). This resulting problem is solved by essentially the same procedure as in the undamped problem considered above. That is, assuming $y(t,x) = w(t)\phi(x)$ and EI and $c_D I$ are constants, we obtain

$$\rho \ddot{w} \phi + EI w \phi'''' + c_D I \dot{w} \phi'''' = 0.$$

Further simplifications yield

$$\frac{\ddot{w}}{w} + \frac{EI}{\rho} \frac{\phi''''}{\phi} + \frac{c_D I}{\rho} \frac{\dot{w}}{w} \frac{\phi''''}{\phi} = 0. \tag{6.44}$$

The above equation (6.44) is not readily separable. However, if we assume that

$$\frac{EI}{\rho} \frac{\phi''''}{\phi} = \beta$$

as in the undamped case above, where β is a positive constant, then we obtain the same mode shape functions as before

$$\phi_n(x) = \sin\left(\frac{n\pi}{L} x\right)$$

and the same separation constants $\beta_n = \frac{EI}{\rho} \left(\frac{n\pi}{L}\right)^4$, for $n = 1, 2, \ldots$. Furthermore, equation (6.44) now becomes

$$\frac{\ddot{w}}{w} + \beta_n + \zeta_n \frac{\dot{w}}{w} = 0, \tag{6.45}$$

where $\zeta_n = \frac{c_D I}{EI} \beta_n$. Equation (6.45) is a simple linear, second-order ordinary differential equation with constant coefficients. Its solution is given by

$$w_n(t) = e^{-\frac{\zeta_n}{2} t} \left(A_n \cos \tilde{\omega}_n t + B_n \sin \tilde{\omega}_n t\right),$$

where $\tilde{\omega}_n = \sqrt{\beta_n - \frac{\zeta_n^2}{4}} = \sqrt{\beta_n \left(1 - \frac{(c_D I)^2}{(EI)^2} \frac{\beta_n}{4}\right)}$. It is worth observing that the undamped frequencies are larger than the damped ones; that is, $\omega_n > \tilde{\omega}_n$ for $n = 1, 2, \ldots$. *Consequently, since all structures and materials have damping, we should expect that the actual frequencies depend on this damping and are lower than the calculated so-called "natural" frequencies which ignore damping in the model.*

In this section we have shown how the method of separation of variables can be used to obtain the solution to beam vibration problems. In general, the method can be extended to a larger class of problems including problems described by more general differential equations, more general boundary conditions, or different geometrical regions. Because of its relative simplicity and wide applicability in many important physical applications, the method of separation of variables remains a method of great importance today. However, this method does have certain limitations. In the first place, the differential equation must be linear so that the principle of superposition can be applied to construct additional solutions by taking linear combinations of the fundamental solutions of the appropriate homogeneous problem. Secondly, in some problems to which the method of separation of variables can be applied in principle, the solvability of the ordinary differential equations is not trivial. For example, in general, EI and/or ρ are not constant but spatially dependent, which would render the method of separation of variables of very limited practical value due to the lack of information about the solutions of the ordinary differential equation (6.26). In the next two sections we will show how a large class of problems can be represented by truncated series similar in spirit to that in equation (6.39). Moreover, we will show that the coefficients in these representations can be determined in a very simple manner.

6.5 Numerical Approximations: Galerkin's Method

In the last section we showed how to solve the simply supported damped beam vibration problem by the method of separation of variables. In principle, we sought solutions of the form

$$y(t, x) = w(t)\phi(x),$$

which led to the representation

$$y(t, x) = \sum_{n=1}^{\infty} w_n(t)\phi_n(x). \tag{6.46}$$

In that case the eigenfunctions or mode shapes are given by

$$\phi_n(x) = \sin\left(\frac{n\pi}{L}x\right)$$

and

$$w_n(t) = w_n(t) = e^{-\frac{\zeta_n}{2}t}\left(A_n \cos \tilde{\omega}_n t + B_n \sin \tilde{\omega}_n t\right),$$

where the coefficients A_n and B_n are determined from the initial conditions (6.21)-(6.22).

It is clear that if we take only a finite number N of terms in the series (6.46), then we obtain only an approximation y_N of y:

$$y(t,x) \approx y_N(t,x) = \sum_{n=1}^{N} w_n(t)\phi_n(x). \tag{6.47}$$

The N basis functions $\{\phi_1, \phi_2, \ldots, \phi_N\}$ define an N-dimensional subspace since each function y_N is determined by a linear combination of only the N functions $\phi_1, \ldots, \phi_N$ in equation (6.47). It should be emphasized that, in general, the basis functions $\phi_i(x)$ need not be trigonometric but may be less smooth functions. In fact, in *finite element methods*, the main idea is that the basis functions ϕ_i are defined piecewise over subregions of the domain called *finite elements*. In addition, over any subdomain, the ϕ_i can be chosen to be very simple functions such as polynomials of low degree.

We are now ready to discuss *Galerkin's* method for constructing approximate solutions to a model problem (equation (6.5) with $\gamma_d = 0$):

$$\rho \ddot{y} + M'' = f \tag{6.48}$$

with boundary condition

$$y(t,0) = y(t,L) = 0,$$

$$M(t,0) = M(t,L) = 0, \tag{6.49}$$

where $M(t,x) = EIy'' + c_D I \dot{y}''$ and L is the length of the beam. The initial condition is given by

$$y(0,x) = y_1(x), \tag{6.50}$$

$$\dot{y}(0,x) = y_2(x). \tag{6.51}$$

We begin by considering our model problem in a *weak* form given as follows: find the function y such that the differential equation, together with the boundary conditions, are satisfied in the sense of weighted averages. Precisely, by the satisfaction of the differential equation in a weighted average sense, we mean that

$$\int_0^L [\rho \ddot{y} + M''] \phi \, dx = \int_0^L f\phi \, dx \tag{6.52}$$

for all members ϕ that belong to a suitable class of functions. In the equation (6.52) the function ϕ is called the *weight function* or *test function* that has to be sufficiently smooth so that the integrals make sense. In fact, the test functions in weak formulations such as (6.52) may not belong to the same class of functions as the class to which the solution y belongs as a function of x. The class of functions to which the solution y belongs as a function of x is called the *class of trial functions*. For example, y might be chosen to be in a class of functions such that their fourth spatial derivatives, when multiplied by a test function ϕ, produce a function $y''''\phi$ that is integrable on the interval $(0, L)$. However, in the weak form (6.52), the test function has no derivatives at all. Hence, although the equation (6.52) is correctly formulated as a variational or weak form of the model problem (6.48), the spaces to which the solution and the test function belong are not the same. Consequently, the weak form (6.52) may not be suitable for easy theoretical or computational considerations.

Therefore, in order to overcome this lack of symmetry in the formulation, we now assume that the solution y and the test function ϕ are sufficiently smooth functions so that we can perform standard integration by parts twice on the moment term to obtain

$$\int_0^L [\rho\ddot{y}\phi + M\phi''] \, dx + M'\phi|_0^L - M\phi'|_0^L = \int_0^L f\phi \, dx. \tag{6.53}$$

To continue with the above formulation, it is important to identify two types of boundary conditions associated with any differential equations. These are called the *natural* and *essential* boundary conditions. After integration by parts, we examine all boundary terms of the variational formulation. From (6.53) we note that the boundary terms involve both the test function ϕ and the dependent variable y. Coefficients of the test function and its derivatives in the boundary expressions are called the *secondary variables*. Specification of secondary variables on the boundary yields the *natural* boundary conditions. Hence, from (6.53), the secondary variables are M' and M, where $M(t, x) = EIy'' + c_D I\ddot{y}''$, and the natural boundary conditions involve specifying M' and M at the boundary points. We also emphasize that secondary variables always have physical meaning and are quantities of interest. In our model problem, the secondary variables M and M' represent bending moment and shear force, respectively.

On the other hand, the dependent variable of the problem, when expressed in the same form as the test function appearing in the boundary terms, is called the *primary variable*. Specifying the primary variable at the boundary points constitutes the *essential* boundary conditions. For the case under consideration, the test function appears as ϕ and ϕ'. Therefore, the dependent variable y and its derivative y' are the primary variables, and evaluations of these variables at the boundary points constitute essential boundary conditions.

With these definitions behind us, we now return to the weak form (6.53). If we assume that the test functions vanish at the boundary endpoints and apply the boundary conditions (6.49), we obtain

$$\int_0^L [\rho\ddot{y}\phi + M\phi''] \, dx = \int_0^L f\phi \, dx, \tag{6.54}$$

for all admissible test functions ϕ. Therefore, the statement (6.53) can be replaced by the following alternative *weak* or *variational* formulation: find

$y(t, \cdot) \in V = H^2 \bigcap H_0^1$ such that

$$\int_0^L [\rho \ddot{y} \phi + EI y'' \phi'' + c_D I \dot{y}'' \phi''] \, dx = \int_0^L f \phi \, dx, \qquad (6.55)$$

for all $\phi \in V$. Here, the class of test and trial functions is defined by $V = H^2 \bigcap H_0^1$ where $H^2(0, L) = \{\phi \in L_2(0, L) : \phi' \in L_2(0, L), \phi'' \in L_2(0, L)\}$ and $H_0^1(0, L) = \{\phi \in L_2(0, L) : \phi' \in L_2(0, L) \text{ with } \phi(0) = \phi(L) = 0\}$. We note that, from (6.55), there is a certain symmetry in the formulation: that is, the same order of derivatives appear in both the test and trial functions. In addition, as we pass from the weak formulation (6.52) to (6.55) we have progressively weakened the smoothness assumptions (in x) on our solution y and, consequently, enlarged the class of functions for which the weak formulation makes sense.

Galerkin's method consists of seeking an approximate solution to the weak form (6.55) in a finite-dimensional subspace V^N of the space V. That is, we seek an approximate solution y^N in $V^N = \text{span}\{B_0^N, B_1^N, \ldots, B_N^N\}$ of the form

$$y^N(t, x) = \sum_{j=0}^N w_j^N(t) B_j^N(x) \qquad (6.56)$$

such that

$$\int_0^L [\rho \ddot{y}^N \phi + EI(y^N)''(\phi^N)'' + c_D I(\dot{y}^N)''(\phi^N)''] \, dx$$

$$= \int_0^L f \phi^N \, dx, \qquad (6.57)$$

for all $\phi^N \in V^N$. It is noted that B_j^N are assumed to be known basis functions, and hence the approximate solution y^N will be completely determined once the coefficient functions w_j^N are found. Furthermore, in Galerkin's method the test functions ϕ^N are chosen to be the same as the basis functions B_j^N. Hence, to determine the specific functions $w_j^N(t)$ that will characterize the approximate solution y^N, we introduce $\phi^N = B_k^N$, $k = 0, 1, \ldots, N$, and the approximation (6.56) into the equation (6.57) to obtain

$$\int_0^L \left[\rho \sum_j \ddot{w}_j^N(t) B_j^N(x) B_k^N(x) + EI \sum_j w_j^N(t)(B_j^N(x))''(B_k^N(x))'' \right.$$

$$\left. + c_D I \sum_j \dot{w}_j^N(t)(B_j^N(x))''(B_k^N(x))'' \right] dx = \int_0^L f(t, x) B_k^N(x) \, dx, (6.58)$$

for $k = 0, 1, \ldots, N$. Interchanging the summation and the integration, we find

$$\sum_{j=0}^{N} \ddot{w}_j^N(t) \int_0^L \rho B_j^N(x) B_k^N(x) \, dx$$

$$+ \sum_{j=0}^{N} \dot{w}_j^N(t) \int_0^L c_D I (B_j^N(x))''(B_k^N(x))'' \, dx$$

$$+ \sum_{j=0}^{N} w_j^N(t) \int_0^L EI(B_j^N(x))''(B_k^N(x))'' \, dx = \int_0^L f(t,x) B_k^N(x) \, dx, \quad (6.59)$$

for $k = 0, 1, \ldots, N$. The structure of the above equation (6.59) is most easily seen by rewriting it in the following vector form

$$\mathcal{M} \frac{d^2}{dt^2} \vec{w}^N(t) + \mathcal{C} \frac{d}{dt} \vec{w}^N(t) + \mathcal{K} \vec{w}^N(t) = \vec{F}^N(t), \quad (6.60)$$

where the vector functions are defined by

$$\vec{w}^N(t) = \left(w_0^N(t), w_1^N(t), \ldots, w_N^N(t) \right)^T,$$

$$\vec{F}^N(t) = \begin{pmatrix} \int_0^L f(t,x) B_0^N(x) \, dx \\ \int_0^L f(t,x) B_1^N(x) \, dx \\ \vdots \\ \int_0^L f(t,x) B_N^N(x) \, dx \end{pmatrix},$$

and the elements of the matrices $\mathcal{M}$, $\mathcal{C}$, and $\mathcal{K}$ are given by

$$(\mathcal{M})_{ij} = \int_0^L \rho B_i^N(x) B_j^N(x) \, dx$$

$$(\mathcal{C})_{ij} = \int_0^L c_D I (B_i^N(x))''(B_j^N(x))'' \, dx$$

$$(\mathcal{K})_{ij} = \int_0^L EI(B_i^N(x))''(B_j^N(x))'' \, dx,$$

for $i, j = 0, 1, \ldots, N$. These matrices $\mathcal{M}$, $\mathcal{C}$ and $\mathcal{K}$ are referred to as the *mass matrix*, the *damping matrix*, and the *stiffness matrix*, respectively. To solve the second order system of ordinary differential equation (6.60) we need the initial conditions. Substituting the approximation (6.56) into the initial conditions (6.51), we have

$$y^N(0, x) = \sum_{j=0}^{N} w_j^N(0) B_j^N(x) \approx y_1(x). \quad (6.61)$$

In general, we should not expect y_1 to lie in V^N, and hence this initial condition can only be satisfied approximately. To do this, we proceed as follows.

Multiplying both sides of the equality of part (6.61) by $B_k^N(x)$ and integrating both sides from 0 to L, we obtain

$$\int_0^L y_1(x) B_k^N(x)\, dx = \sum_{j=0}^{N} w_j^N(0) \int_0^L B_j^N(x) B_k^N(x)\, dx, \qquad (6.62)$$

for $k = 0, 1, \ldots, N$. Equation (6.62) is a linear system of equations for the unknowns $\left(w_0^N(0), w_1^N(0), \ldots, w_N^N(0)\right)$. Similarly, using the other initial conditions $\dot{y}(0, x) = y_2(x)$ we obtain the initial condition for the time derivative $\frac{d}{dt}(\vec{w}^N(t))$ at $t = 0$.

It is important to note that the quality of approximation is completely determined by the choice of the basis functions B_j^N. Once these functions have been chosen, the determination of coefficients w_j^N reduces to a computational matter, which is one of solving for the solution of the ordinary differential equation (6.60). For example, if the basis functions B_j^N are chosen to be the trigonometric function $\sin \frac{j\pi x}{L}$ as in the Fourier series representation, the mass, damping, and stiffness matrices become diagonal matrices and the approximation (6.56) is known as the *modal* approximation. If one chooses the basis functions B_j^N to be *spline* functions, the matrices $\mathcal{M}$, $\mathcal{C}$, and $\mathcal{K}$ become banded matrices. For example, let us partition the domain of our model problem into N finite elements of equal length $h = h^N = L/N$. One set of adequate elements such that $\tilde{B}_j^N \in H^2 \bigcap H_0^1$ is the standard cubic spline given by

$$\tilde{B}_j^N(x) = \frac{1}{h^3} \begin{cases} (x - x_{j-2})^3, & x \in [x_{j-2}, x_{j-1}], \\ h^3 + 3h^2(x - x_{j-1}) \\ \quad +3h(x - x_{j-1})^2 - 3(x - x_{j-1})^3, & x \in [x_{j-1}, x_j], \\ h^3 + 3h^2(x_{j+1} - x) \\ \quad +3h(x_{j+1} - x)^2 - 3(x_{j+1} - x)^3, & x \in [x_j, x_{j+1}], \\ (x_{j+2} - x)^3, & x \in [x_{j+1}, x_{j+2}], \\ 0, & \text{otherwise}, \end{cases} \qquad (6.63)$$

where $x_j = x_j^N = \frac{jL}{N}$, and $j = -1, 0, 1, \ldots, N, N+1$. However, to satisfy the essential boundary conditions $B_j^N(0) = B_j^N(L) = 0$ we further modify these $N + 3$ cubic splines to be of the form

$$B_j^N(x) = \begin{cases} \tilde{B}_0^N(x) - 2\tilde{B}_{-1}^N(x) - 2\tilde{B}_1^N(x), & j = 0, \\ \tilde{B}_j^N(x), & j = 1, 2, \ldots, N-1 \\ \tilde{B}_N^N(x) - 2\tilde{B}_{N-1}^N(x) - 2\tilde{B}_{N+1}^N(x), & j = N. \end{cases}$$

The resulting mass, damping, and stiffness matrices in the finite-dimensional system (6.60) are 7 bandwidth banded matrices with upper and lower bandwidth equal to 3. The space $V^N = \text{span}\{B_j^N\}$ is then of dimension $N + 1$.

6.6 Energy Functional Formulation

In the previous section, we refer to our weak form (6.54) as variational form. This reference arises from the fact that whenever the operators involved possess a certain symmetry, a weak form of the problem can be obtained from a standard problem in the calculus of variations. In such cases, the variational boundary-value problem represents a characterization of the function y that minimizes the energy of the problem.

To illustrate, we now recall the strong form of the undamped, simply supported beam vibration problem

$$\rho \ddot{y} + EI y'''' = 0 \tag{6.64}$$

with boundary conditions

$$y(t,0) = y(t,L) = 0,$$

$$EI y''(t,0) = EI y''(t,L) = 0, \tag{6.65}$$

and initial conditions

$$y(0,x) = y_1(x), \tag{6.66}$$
$$\dot{y}(0,x) = y_2(x), \tag{6.67}$$

where, without loss of generality, we assume that EI is constant.

The associated kinetic energy and potential energy are given by

$$\mathcal{E}_k(t) = \frac{1}{2} \int_0^L \rho \dot{y}(t,x)^2 \, dx, \tag{6.68}$$

$$\mathcal{E}_p(t) = \frac{1}{2} \int_0^L EI y''(t,x)^2 \, dx, \tag{6.69}$$

respectively. We also denote the *action* to be the real-valued function given by

$$A[y](t_0, t_1) = \int_{t_0}^{t_1} (\mathcal{E}_k - \mathcal{E}_p) \, dt. \tag{6.70}$$

Note that A is a "function of functions" and the values of A are real. Any function with these properties is termed a *functional*. Using Hamilton's Principle of Least (Stationary) Action, which states that "On any interval $[t_0, t_1]$, motion (solution of the dynamical problem) provides a stationary value for A," we can consider the classical minimization problem in the calculus of variations. Toward this end, we consider perturbations (also called variations) in the *motion* of the beam y such that

$$y^\epsilon = y + \epsilon \Phi \tag{6.71}$$

is also an admissible motion. Here, the variations $\Phi(t, x) = \eta(t)\phi(x)$ are such that y^ϵ satisfies essential boundary conditions. Hence, $\phi(0) = \phi(L) = 0$. Moreover, for arbitrary t_0 and t_1, we must have

$$y^\epsilon(t_0, x) = y(t_0, x),$$
$$y^\epsilon(t_1, x) = y(t_1, x),$$

which imply that $\eta(t_0) = \eta_1(t_1) = 0$. Finally, the functions η and ϕ must be sufficiently smooth so that the integral in the functional A makes sense. Consequently, we choose $\eta \in C^1(t_0, t_1)$ and assume that ϕ belongs to our previously defined class $H^2 \bigcap H_0^1$.

Using Hamilton's Principle of Least Action, we find the motion y provides a minimum (stationary point) to $A[y + \epsilon\Phi]$ at $\epsilon = 0$. Hence, from the first order necessary condition of optimality,

$$\frac{d}{d\epsilon} A[y + \epsilon\Phi]\big|_{\epsilon=0} = 0,$$

where

$$A[y + \epsilon\Phi] = \int_{t_0}^{t_1} \left[\frac{1}{2} \int_0^L \rho(\dot{y} + \epsilon\dot{\Phi})^2 \, dx - \frac{1}{2} \int_0^L EI(y'' + \epsilon\Phi'')^2 \, dx \right] dt.$$

We thus obtain

$$\frac{d}{d\epsilon} A[y + \epsilon\Phi]\big|_{\epsilon=0} = \int_{t_0}^{t_1} \int_0^L [\rho\dot{y}\dot{\Phi} - EIy''\Phi''] \, dxdt = 0, \tag{6.72}$$

for all admissible variations $\Phi = \eta\phi$. Introducing this substitution into equation (6.72) and integrating by parts, we find

$$\int_{t_0}^{t_1} \int_0^L [\rho\dot{y}\dot{\Phi} - EIy''\Phi''] \, dxdt =$$

$$-\int_{t_0}^{t_1} \int_0^L \rho\ddot{y}\eta\phi \, dxdt + \int_0^L [\rho\dot{y}\eta\phi] \, dx\big|_{t_0}^{t_1} - \int_{t_0}^{t_1} \int_0^L EIy''\eta\phi'' \, dxdt =$$

$$-\int_{t_0}^{t_1} \int_0^L [\rho\ddot{y}\phi + EIy''\phi'']\eta \, dxdt = 0,$$

where we used $\eta(t_0) = \eta(t_1) = 0$. Since η are arbitrary C^1 functions, we must have for all $\phi \in H^2 \bigcap H_0^1$

$$\int_0^L [\rho\ddot{y}\phi + EIy''\phi''] \, dx = 0, \tag{6.73}$$

which is our previously considered weak formulation (6.55) with $c_D I = 0$ and $f = 0$. Hence, we see that the first order necessary condition of variational theory is the same as the weak form of the beam equation. This observation provides the rationale for our use of the term *"variational"* formulation when we refer to the weak formulation (6.73) of the beam problem (6.64).

6.7   The Finite Element Method

In this section, we formally give an introduction to the finite element method in one space dimension, which was discussed previously in the context of the beam problem. In essence, the finite element method is a general and systematic technique for constructing approximate solutions to boundary-value problems. The method involves the application of variational concepts to construct an approximation of the solution over the collection of *finite elements*. The method has been shown to be successful in solving a wide range of problems in engineering and science. Here we give an introduction in the context of second order (in space) systems.

From a mathematical point of view, a convenient way to introduce the finite element method is through the *method of weighted residuals* (see, e.g., [11, 15]). To this end, we consider the following initial boundary value model problem in one-spatial dimension:

$$u_t = \mathcal{L}u(t, x), \qquad t > 0, \qquad x \in [0, L], \qquad (6.74)$$

with initial condition

$$u(0, x) = u_0(x),$$

and boundary conditions

$$u(t, 0) = f_1(t),$$

$$u(t, L) = f_2(t).$$

In the model equation (6.74), $\mathcal{L}$ denotes a *second order spatial* differential operator. For example, in the case that $\mathcal{L}u = \frac{\partial^2}{\partial x^2} u(t, x)$, then the model equation (6.74) becomes the well-known one-dimensional heat equation (see Chapter 5).

To approximate the solution to the model equation (6.74) we begin by subdividing the one-dimensional spatial domain $[0, L]$ into N equally subintervals, also called *finite elements*. Within each element, certain points are identified, called *nodes* or *nodal points*, which play an important role in finite element construction. In the case of equal length finite elements, the nodes are defined by

$$x_i = \frac{(i - 1)L}{N}, \qquad i = 1, 2, \ldots, N + 1.$$

The collection of elements and nodes, which make up the domain of the approximate problem, is often referred to as a *finite element mesh*.

After constructing the finite element mesh for our model problem, we next proceed to construct a corresponding set of basis functions B_i. The basis functions are defined so that the approximate solution $u_A(t, x)$ to our model equation is represented by

$$u_A(t, x) = \sum_{i=1}^{N+1} c_i(t) B_i(x).$$

In addition, these basis functions usually are required to satisfy the following constraints:

1. The basis functions are simple, piecewise functions defined over the finite element mesh;

2. The basis functions are smooth enough so that the approximate solution makes sense in some appropriate space;

3. The basis functions are chosen in such a way so that the coefficient functions $c_i(t)$ are precisely the values of u_A at the nodal points.

One very simple set of basis functions, which are called *hat functions*, is depicted in Figure 6.26. These functions are given by

$$B_i(x) = \begin{cases} \frac{x - x_{i-1}}{x_i - x_{i-1}}, & x \in [x_{i-1}, x_i], \\ \\ \frac{x_{i+1} - x}{x_{i+1} - x_i}, & x \in [x_i, x_{i+1}]. \end{cases}$$

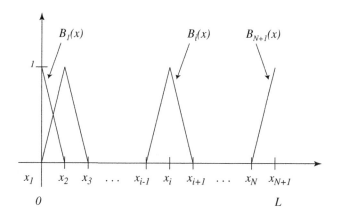

FIGURE 6.26: Hat basis functions.

More generally, one selects basis function $B_i(x)$ so that the approximate solution u_A is of the form

$$u_A(t, x) = u_B(t, x) + \sum_{i=1}^{N+1} c_i(t) B_i(x),$$

where

$$u_B(t, 0) = f_1(t) \text{ and } u_B(t, L) = f_2(t)$$

and $B_i(x) = 0$ for $x = 0, L$. Therefore, $u_A(t, x)$ satisfies the boundary conditions but not the initial condition or the differential equation. This is called the *interior method*. On the other hand, if $u_A(t, x)$ is chosen to satisfy the differential equation but not the boundary condition, this is known as the *boundary method*. For the *mixed method*, u_A satisfies neither the differential equation nor the boundary condition.

We next form the residuals

$$R_E(u_A) = \mathcal{L}u_A - (u_A)_t$$
$$R_I(u_A) = u_0(x) - u_A(0, x).$$

If u_A is the exact solution, both residuals are zero. In the *weighted residual method* (WRM) [11, 15], the coefficient functions $c_i(t)$ are chosen in such a way that the residuals are zero in some "average sense." That is, we choose the weight function $\tilde{\phi}(x)$ so that

$$\int_0^L \tilde{\phi}(x) R_E(u_A(t, x)) dx = 0.$$

However, this gives us only one equation for $N + 1$ unknowns $c_i(t)$. To obtain $N + 1$ equations, we choose $N + 1$ different weighting functions $\tilde{\phi}_j(x)$, $j = 1, 2, \ldots, N + 1$ so that

$$\int_0^L \tilde{\phi}_j(x) R_E(u_A(t, x)) dx = 0, \quad \text{for } j = 1, 2, \ldots, N + 1$$

and

$$\int_0^L \tilde{\phi}_j(x) R_I(u_A(0, x)) dx = 0.$$

Weighted residual methods differ from one another through the choice of $\tilde{\phi}_j(x)$.

- **Galerkin Method**

  Perhaps the best known of these approximate methods is the Galerkin method discussed above (sometimes referred to as Bubnov-Galerkin procedure). Here, the weighting functions $\tilde{\phi}_j(x)$ are chosen to be the same as the basis functions $B_j(x)$,

  $$\tilde{\phi}_j(x) = B_j(x).$$

- **Least Squares Method**

  Let

  $$I(\vec{c}) = \int_0^L R_E^2 dx,$$

  where

  $$\vec{c} = \begin{pmatrix} c_1 \\ c_2 \\ \vdots \\ c_{N+1} \end{pmatrix}.$$

  The idea behind the least squares method is to find a stationary point of $I(\vec{c})$. That is, we seek c_i so that $I(\vec{c})$ is minimized. A necessary optimality condition is given by

  $$\frac{\partial I}{\partial c_j} = 2 \int_0^L R_E \frac{\partial R_E}{\partial c_j} dx = 0, \qquad j = 1, 2, \ldots, N+1,$$

  where $\frac{\partial R_E}{\partial c_j} = \tilde{\phi}_j(x)$. These equations provide $N+1$ equations to solve for $N+1$ unknown functions c_j.

- **Collocation**

  In this method, we select $N+1$ nodal points x_j, $j = 1, 2, \ldots, N+1$ and the corresponding weighting functions

  $$\tilde{\phi}_j(x) = \delta(x - x_j),$$

  where δ is the Dirac delta generalized function. Then

  $$\int_0^L \tilde{\phi}_j(x) R_E(u_A(t, x)) w_j dx = \int_0^L \delta(x - x_j) R_E(u_A(t, x)) dx$$
  $$= R_E(u_A(t, x_j))$$
  $$= 0.$$

  This equation specifies that the residual is zero at $N+1$ specified nodal points x_j. One possible choice would be to choose $B_j(x)$ to be the Chebyshev polynomials and use the roots of a Chebyshev polynomial as the collocation points. The Chebyshev polynomials are defined recursively by

  $$P_0 = 1, \qquad P_1 = x, \qquad P_2 = 2x^2 - 1,$$

  $$P_{r+1}(x) = 2x P_r(x) - P_{r-1}(x),$$

  for $-1 \le x \le 1$. This procedure is sometimes referred to as *orthogonal collocation* in the literature.

6.8   Experimental Beam Vibration Analysis

In this section we describe a physical experiment that is routinely carried out by students in our laboratory (see http://www.ncsu.edu/crsc/ilfum.html). It can be used to perform modal analysis as well as model validation for a cantilever beam. The general arrangement of the hardware needed to setup this experiment is depicted in Figure 6.27. In particular, our experimental setup involves a cantilever beam in a "smart material" paradigm. One end of the beam is clamped while the other end is free, and the beam is mounted with two self-sensing, self-actuating piezoceramic patches. Piezoceramic patches are made up of lead zirconate titanates (PZT's), a piezoelectric material. This type of material belongs to a class of dielectrics that exhibits significant material deformations in response to an applied electric field, and produces dielectric polarization in response to mechanical strains. Therefore, piezoelectric materials have actuating as well as sensing capabilities. The beam in our laboratory can be excited by two sources: (a) an impulse excitation (by a hammer hit) (see Figure 6.27) and (b) a transient periodic excitation (through piezoceramic actuators). The beam transverse acceleration is measured by accelerometers which are attached to the beam structure. In addition, the beam transverse displacement is measured by a proximity transducer system. The proximity transducer system is a gap-to-voltage transducer system that provides accurate, noncontacting static as well as dynamic displacement measurements. Data are recorded and analyzed by the four-channel Hewlett-Packard (HP) dynamic signal analyzer. The HP analyzer allows for both time domain and frequency domain analysis. Finally, in addition to providing data for model validation and modal analysis, this experimental setup has also been used by our students in mechanical vibrational control studies (see Chapter 7). More specifically, for the control of transient vibrations we also added a rapid control prototyping (RCP) system. This system consists of 450MHz Pentium II PC, a digital signal processor (DSP) controller board, a multi-channel filter instrument, and a multi-channel amplifier. The DSP board, a DS1103 made by dSPACE, is equipped with a PowerPC 604e processor running 333 MHz, 36 ADC Channels, and 8 DAC channels. The DS1103 is supported by dSPACE's Total Development Environment (TDE). Within the TDE, programming is done easily via The MathWorks' Simulink. Its graphical user interface allows the user to design new controller with graphical blocks, instead of hardcoding. The amplifier is needed to amplify the low output voltage from the DSP to drive the piezoceramic patches, and the low-pass filter is used to minimize the effects of aliasing. The interested reader is referred to [12] for further details regarding the experimental setup for beam vibration control.

To carry out the experiments outlined above, the hardware and software, which are listed in Tables 6.2 and 6.3, respectively, are recommended.

TABLE 6.2:　Hardware equipment for beam vibration experiment.

Descriptive name	Probable brand (model)
• Proximity Transducer System	Bently Nevada (7200)
• PiezoBeam Accelerometer	Kistler Instrument Corp. 8630C5
• HP Dynamic Signal Analyzer	Hewlett-Packard 35670A
• Impact Hammer Kit	W.A. Brown Instr. GK291C03
• 4-Channel Transducer Coupler	Kistler Instrument Corp. 5134
• Piezoceramic Patches	EDO Corporation
• Wideband Amplifier	Krohn-Hite Corp. (7600)
• Low-Pass Filter	Frequency Devices (9002)
• DSP Board	dSPACE (1103)

TABLE 6.3:　Software tools for beam vibration experiment.

Descriptive name	Brand
• Real-Time Workshop (for control)	The MathWorks, Inc.
• Simulink (for control)	The MathWorks, Inc.
• MATLAB and Optimization toolbox for data and model analysis	The MathWorks, Inc.

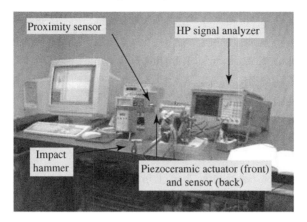

FIGURE 6.27: Hardware used for modal analysis and model validation of the cantilever beam model.

Project: Beam Vibration Analysis

The aim of this project is to carry out modal analysis of the mathematical model for the transverse displacement of a cantilever beam. In addition, the second part of the project involves a parameter estimation problem using data collected from the experiment as described in §6.8. For readers without accessibility to such experimental data, simulation data from a numerical (e.g., finite element) solution with noise added can be generated and used in the place of experimental observations.

A. Mode Shapes and Natural Frequencies

Consider a cantilever beam with length, $L = 1$, and free end at $x = L$. Assume that ρ and EI are constants with their values equal to 1 and 5 respectively. The free-vibration equation of motion for this system is

$$\rho\frac{\partial^2 y(t,x)}{\partial t^2} + EI\frac{\partial^4 y(t,x)}{\partial x^4} = 0$$

with initial conditions:

$$y(0,x) = y_1(x)$$
$$\frac{\partial y(0,x)}{\partial t} = y_2(x),$$

where y_1 and y_2 are some given functions of x and $x \in [0,1]$. The boundary conditions for this case are given by:

$$y(t,0) = 0, \qquad \frac{\partial y(t,0)}{\partial x} = 0$$

$$M(t,1) = EI\frac{\partial^2 y(t,1)}{\partial x^2} = 0, \qquad \frac{\partial M(t,1)}{\partial x} = EI\frac{\partial^3 y(t,1)}{\partial x^3} = 0.$$

1.) Using separation of variables, that is, letting

$$y(t,x) = w(t)\phi(x),$$

show that

$$\phi(x) = A_1 \cos(\xi x) + A_2 \sin(\xi x) + A_3 \cosh(\xi x) + A_4 \sinh(\xi x),$$

where A_i are constants, $\xi = (\beta\rho/EI)^{1/4}$ and β is a constant to be determined.

2.) Applying the boundary conditions, show that

$$A_3 = -A_1, \qquad A_4 = -A_2,$$

and ξ satisfies $\cos(\xi) = -(1/\cosh(\xi))$.

3.) Plot the functions $\cos(\xi)$ and $-(1/\cosh(\xi))$ on the same graph for $\xi \in [0,50]$. Note that the intersections of the two plots give the values of β and there are infinitely many such values.

4.) Estimate the first five values of β from the graph in part 3.) and use these values as estimates for the MATLAB routine `fzero` to compute the first five zeros of the function $\cos(\xi) + (1/\cosh(\xi)) = 0$. Show that the fourth and higher zeros can be approximated accurately (within four decimal places) by

$$\xi_i = \frac{\pi}{2}(2i - 1), \qquad i = 4, 5, 6, \ldots,$$

which are the zeros of $\cos x$.

5.) Using these computed values of ξ from part 4.) compute the first five natural frequencies of the beam ω_i, $i = 1, 2, 3, 4, 5$ and plot the first five mode shape functions $\phi_i(x)$, $i = 1, 2, 3, 4, 5$.

6.) Now consider the same cantilever beam as above but with viscous damping of the form $\gamma\frac{\partial y}{\partial t}$ where γ is a constant and is equal to 2. How are the frequencies and mode shapes different from the undamped case? If they are different, compute the first five frequencies and mode shape functions.

B. Experimental Data

Consider the same cantilever beam as in part A. but with length L being the length of the beam which was used in the experiments.

1.) Suppose that the parameters ρ and EI are unknown but the first few natural frequencies of the beam, ω_i, are known (for example, from the physical experiments that you performed). Discuss how one can use the natural frequencies to compute the unknown parameters ρ and EI.

2.) From the experiments performed in the laboratory (or from simulated data), you should have at least two data sets. One data set contains the displacement of the beam as a function of time. The other data set contains the power spectrum data whose graph provides information on the natural frequencies of the beam (that is, the location of the peaks in the plot indicates the frequency components present in the signal).

    (a) Perform the discrete Fourier transform (DFT) on your time domain data set and verify that the location of peaks in the plot of the DFT correspond to those in the power spectrum data set. Recall that, for real data, the N-point DFT is symmetric around the $N/2$ point, so for plotting purposes, it is sufficient to plot the first half of the DFT, which corresponds to positive frequencies.

    (b) How many frequency components do you observe in the signal? What are their values?

    (c) Using the frequency information, estimate the unknown parameters ρ and EI of the beam.

    (d) Combining direct computation (for example, the moment of inertia, I, can be computed exactly from the known dimensions of the beam) with a literature search, find the "book" values of ρ and EI for the beam that you used in the experiments. How are these values different from those that you computed in part (c)? Explain what you think are the reasons for the differences.

References

[1] H.T. Banks, W. Fang, R.J. Silcox and R.C. Smith, Approximation methods for control of acoustic/structure models with piezoceramic actuators, *Journal of Intelligent Material Systems and Structures*, **4**(1), 1993, pp. 98–116.

[2] H.T. Banks, N.G. Medhin and G.A. Pinter, Multiscale considerations in modeling of nonlinear elastomers, CRSC-TR03-42, North Carolina State University, Raleigh, North Carolina, October, 2003; *Journal of Computational Methods in Science and Engineering*, **8**, 2007, pp. 53–62.

[3] H.T. Banks, R.J. Silcox and R.C. Smith, The modeling and control of acoustic/structure interaction problems via piezoceramic actuators: 2-D numerical examples, *ASME Journal of Vibration and Acoustics*, **116**(3), 1994, pp. 386–396.

[4] W.E. Boyce and R.C. DiPrima, *Elementary Differential Equations and Boundary Value Problems*, John Wiley & Sons, Inc., Hoboken, 8th ed., 2004.

[5] A.J. Bullmore, P.A. Nelson, A.R.D. Curtis and S.J. Elliott, The active minimization of harmonic enclosed sound fields, part II: A computer simulation, *Journal of Sound and Vibration*, **117**(1), 1987, pp. 15–33.

[6] R. Courant and D. Hilbert, *Methods of Mathematical Physics, vol. II*, Wiley, New York, 1962.

[7] J.M. Gere and S.P. Timoshenko, *Mechanics of Materials*, PWS Pub. Co., Boston, 4th ed., 1997.

[8] W. Kaplan, *Advanced Calculus*, Addison-Wesley Publishing Co., Inc., New York, 1991.

[9] H.C. Lester and C.R. Fuller, Active control of propeller induced noise fields inside a flexible cylinder, *AIAA 10th Aeroacoustic Conference*, Seattle, WA, 1986.

[10] E.P. Popov, *Introduction to Mechanics of Solids*, Prentice-Hall, Inc., Englewood Cliffs, 1968.

[11] J.N. Reddy, *An Introduction to the Finite Element Method*, McGraw Hill Series in Mechanical Engineering, New York, 3rd ed., 2005.

[12] R.C.H. del Rosario, H.T. Tran and H.T. Banks, Proper orthogonal decomposition based control of transverse beam vibrations: Experimental implementation, CRSC-TR99-43, North Carolina State University, Raleigh, North Carolina, 1999; *IEEE Trans. on Control Systems Technology*, **10**, 2002, pp. 717–726.

[13] K.R. Symon, *Mechanics*, Addison-Wesley Publishing Co., Reading, 1971.

[14] S.P. Timoshenko and J.N. Goodier, *Theory of Elasticity*, McGraw-Hill, Inc., New York, 1987.

[15] O.C. Zienkiewicz and R.C. Taylor, *Finite Element Method: Volume 1, The Basis*, Butterworth-Heinemann, Newbury (UK), 5th ed., 2000.

Chapter 7

Beam Vibrational Control and Real-Time Implementation

7.1 Introduction

In this chapter we focus on the real-time implementation of feedback controls for the attenuation of transverse beam vibrations due to transient pulsation. In particular, we will consider an aluminum cantilever beam to which two piezoceramics patches are mounted in a symmetric opposing fashion. The sensing device to be used for observation is a proximity probe, and thus the sensor loading effects on the beam (an extremely thin metallic surface mounted on the beam) are assumed to be negligible and are not taken into account in the modeling of the beam. Also, it is assumed that the beam vibration occurs transversely (with no out of plane torsion or twisting about the axis of the beam), a reasonable assumption for beams that have relatively small thickness when compared to width. Hence we can make use of the Euler-Bernoulli beam model that we developed earlier in Chapter 6. Most of the facts presented in this section are standard knowledge and can be found in the textbook literature (see, e.g., [1, 2, 4, 5, 6, 17]).

The control methodology to be discussed in this chapter is the well-known linear quadratic regulator (LQR) design method. We will discuss how to implement this control in real-time where only partial state measurements are available (transverse beam displacement data at a single location on the beam). Such considerations require the use of an observer or state estimator coupled with full state feedback. To illustrate these control methodologies, we will begin by reviewing several important concepts from control theory.

7.2 Controllability and Observability of Linear Systems

In control theory the typical problem is to find the input (or control) that causes the state or the output to behave in a desired way. In particular, two basic questions in control theory are:

(a) Is it possible to find a suitable control input that can transform any initial state to any desired state in a finite length of time?

(b) Is it possible to identify or reconstruct the initial state by observing the output in a finite length of time?

Consider a simple example where the state $(x_1(t), x_2(t))^T$ and the output $y(t)$ of a dynamical system are given by:

$$\begin{pmatrix} \dot{x}_1(t) \\ \dot{x}_2(t) \end{pmatrix} = \begin{pmatrix} -1 & 0 \\ 0 & 2 \end{pmatrix} \begin{pmatrix} x_1(t) \\ x_2(t) \end{pmatrix} + \begin{pmatrix} 1 \\ 0 \end{pmatrix} u(t),$$ (7.1)

$$y(t) = \begin{pmatrix} 0 & 1 \end{pmatrix} \begin{pmatrix} x_1(t) \\ x_2(t) \end{pmatrix}.$$ (7.2)

It is clear that no matter what input $u(t)$ is applied, the state variable $x_2(t)$ is not affected. Hence $x_2(t)$ is said to be not controllable by the input $u(t)$. On the other hand, the state variable $x_2(t)$ can be measured or observed but $x_1(t)$ is neither observable nor measurable. Hence $x_1(t)$ is said to be not observable from the output $y(t)$. This example illustrates the concepts of controllability and observability that we shall explain in more detail below.

Remark 7.2.1 Kalman introduced the ideas of controllability and observability in [11]. Another detailed exposition of these concepts can be found in [12].

7.2.1 Controllability

7.2.1.1 Time-Varying Case

Consider the n-dimensional linear state equation

$$\dot{\vec{x}}(t) = A(t)\vec{x}(t) + B(t)\vec{u}(t),$$ (7.3)

where $\vec{x}(t) \in \mathbb{R}^n, \vec{u}(t) \in \mathbb{R}^m$ and $A(\cdot)$ and $B(\cdot)$ are $n \times n$ and $n \times m$ matrices whose elements are continuous functions on $(-\infty, \infty)$. Because the output does not play a role in controllability, we will not consider the output equation for now.

Definition 7.2.1 *The state system (7.3) is said to be (completely) <u>controllable</u> at time t_0 if for any pair of states $\vec{x}_0$ and $\vec{x}_1 \in \mathbb{R}^n$ there is a finite time $t_1 > t_0$ and an input $\vec{u}(\cdot)$ on $[t_0, t_1]$ such that $\vec{u}(t)$ transfers $\vec{x}_0$ to the state $\vec{x}_1$ at time t_1. That is,*

$$\vec{x}_1 = \Phi(t_1, t_0)\vec{x}_0 + \int_{t_0}^{t_1} \Phi(t_1, s)B(s)\vec{u}(s)ds,$$ (7.4)

where $\Phi(t, t_0)$ is the $n \times n$ state transition matrix satisfying

$$\dot{\Phi}(t, t_0) = A(t)\Phi(t, t_0)$$
$$\Phi(t_0, t_0) = I.$$

Remark 7.2.2

(i) If $A(t)$ has the following commutative property

$$A(t) \left(\int_{t_0}^t A(s)ds \right) = \left(\int_{t_0}^t A(s)ds \right) A(t),$$

then

$$\Phi(t, t_0) = e^{\int_{t_0}^t A(s)ds},$$

and the solution to

$$\dot{\vec{x}}(t) = A(t)\vec{x}(t), \quad \vec{x}(t_0) = \vec{x}_0$$

is given by

$$\vec{x}(t) = \Phi(t, t_0)\vec{x}_0,$$

which is a transformation of the initial condition. For this reason, $\Phi(t, t_0)$ is called the *state transition matrix*. Note that the above commutative condition holds in particular if the elements of A are constants.

(ii) In this definition, the term "completely" means that the definition holds for all $\vec{x}_0$ and $\vec{x}_1$. The control $\vec{u}(\cdot)$ is assumed to be either piecewise continuous on $[t_0, t_1]$ or in $L_2[t_0, t_1]$.

(iii) Rewrite equation (7.4) as

$$\vec{x}_1 - \Phi(t_1, t_0)\vec{x}_0 = \int_{t_0}^{t_1} \Phi(t_1, s)B(s)\vec{u}(s)ds.$$

Controllability means that this equation is solvable for $\vec{u}(t)$ given arbitrary $\vec{x}_1$, $\vec{x}_0$, t_0 and t_1. In addition, if we let $\tilde{x}_1 = \vec{x}_1 - \Phi(t_1, t_0)\vec{x}_0$, then

$$\tilde{x}_1 = \int_{t_0}^{t_1} \Phi(t_1, s)B(s)\vec{u}(s)ds,$$

which shows that $u(\cdot)$ also transfers the state from $\vec{x}_0 = \vec{0}$ to $\tilde{x}_1$ on $[t_0, t_1]$. Hence an equivalent definition of controllability means that every state can be reached from the origin in finite time. On the other hand, if we rewrite equation (7.4) as

$$\vec{0} = \Phi(t_1, t_0)\vec{x}_0 - \vec{x}_1 + \int_{t_0}^{t_1} \Phi(t_1, s)B(s)\vec{u}(s)ds$$

$$= \Phi(t_1, t_0)[\vec{x}_0 - \Phi(t_0, t_1)\vec{x}_1] + \int_{t_0}^{t_1} \Phi(t_1, s)B(s)\vec{u}(s)ds,$$

then the same control $u(\cdot)$ also transfers the state from $\tilde{x}_0 = \vec{x}_0 - \Phi(t_0, t_1)\vec{x}_1$ to $\vec{0}$. That is, controllability also means that every state can be transferred to the origin in finite time.

(iv) This definition requires only that $\vec{u}(t)$ be capable of moving any state in the state space to any other state in finite time; the exact state trajectory is not specified.

(v) Controllability has nothing to do with the output and no constraint is imposed on the input.

(vi) Controllability implies the existence of an open-loop control, but it does not tell us how to construct one.

It is well known that the solution to the state equation (7.3) is given by the variation of constants formula (see, e.g., [6])

$$\vec{x}(t) = \Phi(t, t_0)\vec{x}(t_0) + \int_{t_0}^{t} \Phi(t, s)B(s)\vec{u}(s)ds$$

$$= \Phi(t, t_0)\left[\vec{x}(t_0) + \int_{t_0}^{t} \Phi(t_0, s)B(s)\vec{u}(s)ds\right]. \tag{7.5}$$

Consider the $n \times n$ constant matrix

$$W(t_0, t_1) = \int_{t_0}^{t_1} \Phi(t_0, s)B(s)B^T(s)\Phi^T(t_0, s)ds.$$

Let $\vec{x}(t_0) = \vec{x}_0$ and $\vec{x}_1$ be arbitrary. Assume that $W(t_0, t_1)$ is nonsingular and consider

$$\vec{u}(t) = -B^T(t)\Phi^T(t_0, t)W^{-1}(t_0, t_1)[\vec{x}_0 - \Phi(t_0, t_1)\vec{x}_1].$$

Substituting the above equation for $\vec{u}(t)$ into equation (7.5) we have

$$\vec{x}(t_1) = \Phi(t_1, t_0)\{\vec{x}_0$$
$$- \int_{t_0}^{t_1} \Phi(t_0, s)B(s)B^T(s)\Phi^T(t_0, s)W^{-1}(t_0, t_1)[\vec{x}_0 - \Phi(t_0, t_1)\vec{x}_1]ds\}$$
$$= \Phi(t_1, t_0)\left\{\vec{x}_0 - W(t_0, t_1)W^{-1}(t_0, t_1)[\vec{x}_0 - \Phi(t_0, t_1)\vec{x}_1]\right\}$$
$$= \Phi(t_1, t_0)\Phi(t_0, t_1)\vec{x}_1$$
$$= \vec{x}_1.$$

Thus, $W(t_0, t_1)$ being nonsingular is a sufficient condition for controllability of system (7.3). Equivalently [6], the linearly independence of the rows of $\Phi(t_0, \cdot)B(\cdot)$ on $[t_0, t_1]$ is a sufficient condition for the controllability of (7.3). In fact, the linear independency of the rows of the $n \times m$ matrix $\Phi(t_0, \cdot)B(\cdot)$ on $[t_0, t_1]$ is a necessary and sufficient condition for the controllability of (7.3).

To see the necessary condition, we assume that (7.3) is controllable but the rows of $\Phi(t_0, \cdot)B(\cdot)$ are linearly dependent on $[t_0, t_1]$ for all $t_1 > t_0$. Then there exists a nonzero vector $\vec{c} \in \mathbb{R}^n$ such that

$$\vec{c}^T\Phi(t_0, t)B(t) = 0$$

for all $t \in [t_0, t_1]$. Let us choose $\vec{c} = \vec{x}(t_0) = \vec{x_0}$. Then equation (7.5) can be rewritten as

$$\vec{c}^T \Phi(t_0, t_1)\vec{x}(t_1) = \vec{c}^T\vec{c} + \vec{c}^T \int_{t_0}^{t_1} \Phi(t_0, s)B(s)\vec{u}(s)ds.$$

Since (7.3) is controllable at t_0, for any state $\vec{x}(t_1)$ (in particular for $\vec{x}(t_1) = 0$) there exists a control $\vec{u}(t)$ such that

$$\vec{c}^T\vec{c} = 0,$$

where we used the fact that $\vec{c}^T\Phi(t_0, s)B(s) = 0$ for all $s \in [t_0, t_1]$. This implies that $\vec{c} = \vec{0}$ which is a contradiction.

Remark 7.2.3

(a) The above discussion on controllability involves the computation of the state transition matrix $\Phi(t_0, \cdot)$ which is a very difficult task in general.

(b) The $n \times n$ constant matrix

$$W(t_0, t_1) = \int_{t_0}^{t_1} \Phi(t_0, s)B(s)B^T(s)\Phi^T(t_0, s)ds$$

is called the (controllability) Grammian matrix. From [6], it can be shown that the equivalent necessary and sufficient condition for (7.3) to be controllable at time t_0 is for $W(t_0, t_1)$ to be nonsingular.

Recall that controllability means that there exists a control $\vec{u}(\cdot)$ capable of moving any state in the state space to any other state in finite time. Since the state trajectory is not specified (see Remark 7.2.2(iv)), there are, in general, many different controls $\vec{u}(\cdot)$ that achieve this task. One possible control $\vec{u}(\cdot)$ is given by the formula

$$\vec{u}(t) = -B^T(t)\Phi^T(t_0, t)W^{-1}(t_0, t_1)[\vec{x_0} - \Phi(t_0, t_1)\vec{x_1}] \qquad (7.6)$$

as shown above. If we define the so-called total energy E of (7.3) as

$$E = \int_{t_0}^{t_1} \|\vec{u}(t)\|^2 dt,$$

where $\|\cdot\|$ denotes the Euclidean norm, then the control $\vec{u}(\cdot)$ in equation (7.6) is the one which minimizes this energy and is called the minimum-energy control. That is, if we let $\vec{u}(t)$ be given by (7.6) which transfers $\vec{x_0}$ to $\vec{x_1}$ at time t_1 and let $\tilde{u}(t)$ be another control on $[t_0, t_1]$ that accomplishes the same task, then

$$\int_{t_0}^{t_1} \|\vec{u}(t)\|^2 dt \leq \int_{t_0}^{t_1} \|\tilde{u}(t)\|^2 dt.$$

To see this, we recall that the solution to (7.3) with the initial condition $\vec{x}(t_0) = \vec{x_0}$ is given by

$$\vec{x}(t) = \Phi(t, t_0) \left[\vec{x_0} + \int_{t_0}^{t} \Phi(t_0, s) B(s) \vec{u}(s) ds \right].$$

Letting $\vec{x_1} = \vec{x}(t_1)$, we can rewrite the above equation as

$$\Phi(t_0, t_1) \vec{x_1} - \vec{x_0} = \int_{t_0}^{t_1} \Phi(t_0, s) B(s) \vec{u}(s) ds$$

or

$$\tilde{x}_1 = \int_{t_0}^{t_1} \Phi(t_0, s) B(s) \vec{u}(s) ds,$$

where $\tilde{x}_1 = \Phi(t_0, t_1)\vec{x_1} - \vec{x_0}$. Since $\vec{u}(\cdot)$ and $\tilde{u}(\cdot)$ both transfer $\vec{x_0}$ to $\vec{x_1}$ at t_1, we have

$$\int_{t_0}^{t_1} \Phi(t_0, s) B(s) \vec{u}(s) ds = \int_{t_0}^{t_1} \Phi(t_0, s) B(s) \tilde{u}(s) ds$$

or

$$\int_{t_0}^{t_1} \Phi(t_0, s) B(s) [\vec{u}(s) - \tilde{u}(s)] ds = \vec{0}.$$

This implies that

$$\left\langle \int_{t_0}^{t_1} \Phi(t_0, s) B(s) [\vec{u}(s) - \tilde{u}(s)] ds, W^{-1}(t_0, t_1)\tilde{x}_1 \right\rangle = 0,$$

or equivalently,

$$\int_{t_0}^{t_1} \left\langle \vec{u}(s) - \tilde{u}(s), B^T(s) \Phi^T(t_0, s) W^{-1}(t_0, t_1)\tilde{x}_1 \right\rangle ds = 0.$$

Using (7.6), we find from this equation

$$\int_{t_0}^{t_1} \left\langle \vec{u}(s) - \tilde{u}(s), \vec{u}(s) \right\rangle ds = 0.$$

Then

$$\int_{t_0}^{t_1} \|\tilde{u}(t)\|^2 dt = \int_{t_0}^{t_1} \|\tilde{u}(t) - \vec{u}(t) + \vec{u}(t)\|^2 dt$$

$$= \int_{t_0}^{t_1} \|\tilde{u}(t) - \vec{u}(t)\|^2 dt + \int_{t_0}^{t_1} \|\vec{u}(t)\|^2 dt + 2 \int_{t_0}^{t_1} \left\langle \tilde{u}(t) - \vec{u}(t), \vec{u}(t) \right\rangle dt$$

$$= \int_{t_0}^{t_1} \|\tilde{u}(t) - \vec{u}(t)\|^2 dt + \int_{t_0}^{t_1} \|\vec{u}(t)\|^2 dt.$$

Hence

$$\int_{t_0}^{t_1} \|\tilde{u}(t)\|^2 dt - \int_{t_0}^{t_1} \|\vec{u}(t)\|^2 dt = \int_{t_0}^{t_1} \|\tilde{u}(t) - \vec{u}(t)\|^2 dt$$
$$\geq 0,$$

or

$$\int_{t_0}^{t_1} \|\vec{u}(t)\|^2 dt \leq \int_{t_0}^{t_1} \|\tilde{u}(t)\|^2 dt.$$

We now give a controllability criterion based solely on the system matrices $A(\cdot)$ and $B(\cdot)$. To this end, we assume that $A(\cdot)$ and $B(\cdot)$ are $(n-1)$ times continuously differentiable and let $\Psi(t)$ denote the *fundamental matrix* of $\dot{\vec{x}}(t) = A(t)\vec{x}(t)$. The relationship between the state transition matrix Φ and the fundamental matrix Ψ is given by [6]

$$\Phi(t, t_0) = \Psi(t)\Psi^{-1}(t_0),$$

for all $t, t_0 \in (-\infty, \infty)$. Define a sequence of $n \times m$ matrices $M_k(t)$ by the equations

$$M_0(t) = B(t),$$
$$M_{k+1}(t) = -A(t)M_k(t) + \frac{d}{dt}M_k(t),$$

for $k = 0, 1, \ldots, n-1$. We next observe that

$$\Phi(t_0, t)B(t) = \Phi(t_0, t)M_0(t),$$

$$\frac{\partial}{\partial t}\{\Phi(t_0, t)B(t)\} = \frac{d}{dt}\{\Psi(t_0)\Psi^{-1}(t)B(t)\}$$

$$= \Psi(t_0)\frac{d}{dt}[\Psi^{-1}(t)]B(t) + \Psi(t_0)\Psi^{-1}(t)\frac{d}{dt}B(t)$$

$$= \Psi(t_0)\Psi^{-1}(t)\left[-A(t)B(t) + \frac{d}{dt}B(t)\right],$$

where we have used $\Psi(t)\Psi^{-1}(t) = I(t)$. After differentiating both sides, we obtain

$$\frac{d}{dt}\Psi^{-1}(t) = -\Psi^{-1}(t)A(t).$$

Therefore,

$$\frac{\partial}{\partial t}\{\Phi(t_0, t)B(t)\} = \Phi(t_0, t)M_1(t).$$

Similarly,

$$\frac{\partial^2}{\partial t^2}\{\Phi(t_0, t)B(t)\} = \frac{\partial}{\partial t}[\Phi(t_0, t)]M_1(t) + \Phi(t_0, t)\frac{d}{dt}M_1(t)$$

$$= \Phi(t_0, t)\left[-A(t)M_1(t) + \frac{d}{dt}M_1(t)\right]$$

$$= \Phi(t_0, t)M_2(t).$$

In general,

$$\frac{\partial^k}{\partial t^k}\{\Phi(t_0, t)B(t)\} = \Phi(t_0, t)M_k(t),$$

for $k = 0, 1, \ldots, n-1$. Now consider the matrix

$$\left[\Phi(t_0, t_1)B(t_1)\Big|\frac{\partial}{\partial t}\{\Phi(t_0, t)B(t)\}|_{t=t_1}\Big| \cdots \Big|\frac{\partial^{n-1}}{\partial t^{n-1}}\{\Phi(t_0, t)B(t)\}\Big|_{t=t_1}\right]$$

$$= \Phi(t_0, t_1)[M_0(t_1)|M_1(t_1)|\ldots|M_{n-1}(t_1)]$$

where $t_1 > t_0$. Since $\Phi(t_0, t_1)$ is nonsingular, if

$$\rho([M_0(t_1)|M_1(t_1)|\ldots|M_{n-1}(t_1)]) = n,$$

then

$$\rho\left(\left[\Phi(t_0, t_1)B(t_1)\Big|\frac{d}{dt}\{\Phi(t_0, t_1)B(t_1)\}\Big| \ldots \Big|\frac{\partial^{n-1}}{\partial t^{n-1}}\{\Phi(t_0, t_1)B(t_1)\}\right]\right) = n,$$

(7.7)

where ρ denotes the rank of a matrix and we used the fact that if $A = BC$, $\rho(A) \le \rho(C)$. But then because B is nonsingular, $C = B^{-1}A$ and $\rho(C) \le \rho(A)$. Hence, $\rho(A) = \rho(C)$ if B is nonsingular. The rank condition (7.7) is equivalent to the condition that the rows of $\Phi(t_0, \cdot)B(\cdot)$ are linearly independent on $[t_0, t_1]$ [6]. Hence, a sufficient condition for the controllability of (7.3) at time t_0 is

$$\rho([M_0(t_1)|M_1(t_1)|\ldots|M_{n-1}(t_1)]) = n$$

for some time $t_1 > t_0$.

7.2.1.2 Time-Invariant Case

We now consider the controllability question of the time-invariant state equation

$$\dot{\vec{x}}(t) = A\vec{x}(t) + B\vec{u}(t),$$

(7.8)

where $\vec{x} \in \mathbb{R}^n, \vec{u} \in \mathbb{R}^m$ and A and B are $n \times n$ and $n \times m$ constant matrices, respectively.

In the time-invariant case, $\Phi(t_0, t)B(t) = e^{A(t_0-t)}B$. Elements of the matrix function $e^{A(t_0-t)}$ are of the form $t^k e^{\lambda t}$; hence, elements of $e^{A(t_0-t)}B$ are linear combinations of terms of the form $t^k e^{\lambda t}$ which are analytic on $[0, \infty)$. Consequently, if the rows of $e^{A(t_0-t)}B$ are linearly independent on $[0, \infty)$, they are linearly independent on $[t_0, t_1]$ for any t_0 and $t_1 > t_0$. Hence if the time-invariant system (7.8) is controllable, it is controllable at any time $t_0 \ge 0$. For this reason the reference of t_0 in the definition of controllability will be dropped for the time-invariant system.

In a manner similar to that in the time-variant case discussed in the previous section, one can easily establish the following *equivalence statements* regarding controllability of the time-invariant system of equation (7.8).

(i) The system (7.8) is controllable;

(ii) The n rows of $e^{-At}B$ are linearly independent on $[0, \infty)$;

(iii) The controllability Grammian matrix

$$W(0, t) = \int_0^t e^{-As} B B^T e^{-A^T s} ds \tag{7.9}$$

is nonsingular for any $t > 0$. Furthermore, the control

$$u(t) = -B^T e^{-A^T t} W^{-1}(0, t_1)[\vec{x}_0 - e^{-A^T t_1}\vec{x}_1] \tag{7.10}$$

transfers $\vec{x}(0) = \vec{x}_0$ to $\vec{x}(t_1) = \vec{x}_1$;

(iv) $\rho[Q] = n$ where Q is the $n \times nm$ controllability matrix

$$Q = \begin{bmatrix} B \,|\, AB \,|\, A^2 B \,|\ldots|\, A^{n-1}B \end{bmatrix}. \tag{7.11}$$

Example 7.2.1 In this example, we consider a platform which is supported by springs and dampers as shown in Figure 7.1. This platform system can be used to study suspension systems of automotives. To simplify the mathematical model, we assume that the mass of the platform is negligible. Thus the movements of the two spring systems can be regarded as independent and the applied force u is assumed to be distributed to each spring system as shown. In addition, assume that the dampers are proportional to the velocity and the springs obey Hooke's law. In particular, the viscous damping coefficients of both springs are assumed to be 1 and the spring constants are assumed to be 1 and 2 as depicted in Figure 7.1. If the displacements of the two springs from equilibrium are denoted by x_1 and x_2, then (using the fact that the mass of the platform is very small compared to the spring and damping constants) we obtain from (2.9)

$$x_1 + \dot{x}_1 = u$$
$$2x_2 + \dot{x}_2 = 3u$$

or

$$\begin{pmatrix} \dot{x}_1 \\ \dot{x}_2 \end{pmatrix} = \begin{pmatrix} -1 & 0 \\ 0 & -2 \end{pmatrix} \begin{pmatrix} x_1 \\ x_2 \end{pmatrix} + \begin{pmatrix} 1 \\ 3 \end{pmatrix} u.$$

If the initial displacements of both ends are different from zero, the system will oscillate. Now we pose the following problem: Let $x_1(0) = 10$ and $x_2(0) = -2$. Is it possible to apply a control function $u(t)$ which will bring the system to rest at $t_1 = 1$ second? $t_1 = 0.5$ second?

The controllability matrix, Q, for this platform problem is

$$Q = \begin{pmatrix} 1 & -1 \\ 3 & -6 \end{pmatrix}$$

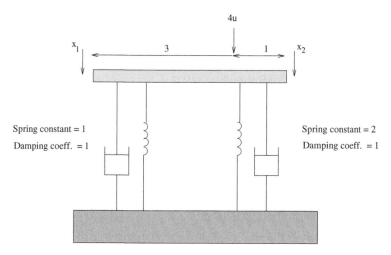

FIGURE 7.1:   A spring-mass-dashpot platform system.

which has rank 2. Hence the system is controllable, and the displacements can be brought to zero in an arbitrarily small time interval from any initial displacements. Using formulas (7.9) and (7.10), we can find the Grammian controllability matrix W and hence the control $\vec{u}(t)$ that will drive the system from the initial state $\vec{x_0}$ to the zero state in $t_1 = 1$ second.

Observe that the matrix A is in Jordan form [6]; therefore,

$$e^{-At} = \begin{pmatrix} e^t & 0 \\ 0 & e^{2t} \end{pmatrix}.$$

Then

$$
\begin{aligned}
W(0,1) &= \int_0^1 \begin{pmatrix} e^s & 0 \\ 0 & e^{2s} \end{pmatrix} \begin{pmatrix} 1 \\ 3 \end{pmatrix} \begin{pmatrix} 1 & 3 \end{pmatrix} \begin{pmatrix} e^s & 0 \\ 0 & e^{2s} \end{pmatrix} ds \\
&= \int_0^1 \begin{pmatrix} e^{2s} & 3e^{3s} \\ 3e^{3s} & 9e^{4s} \end{pmatrix} ds \\
&= \begin{pmatrix} 3.1945 & 19.0855 \\ 19.0855 & 120.5958 \end{pmatrix},
\end{aligned}
$$

and the control $u(t)$ is given by

$$u_{t_1=1}(t) = -\begin{pmatrix} 1 & 3 \end{pmatrix} \begin{pmatrix} e^t & 0 \\ 0 & e^{2t} \end{pmatrix} W^{-1}(0,1) \begin{pmatrix} 10 \\ -2 \end{pmatrix}$$

$$= -59.2751e^t + 28.1925e^{2t}.$$

Similarly, we can find the control that will drive the system to zero in $t_1 = .5$

second. In this case, the controllability Grammian matrix is

$$W(0, .5) = \int_0^{.5} \begin{pmatrix} e^s & 0 \\ 0 & e^{2s} \end{pmatrix} \begin{pmatrix} 1 \\ 3 \end{pmatrix} (1\ 3) \begin{pmatrix} e^s & 0 \\ 0 & e^{2s} \end{pmatrix} ds$$

$$= \begin{pmatrix} 0.8591 & 3.4817 \\ 3.4817 & 14.3754 \end{pmatrix},$$

and the control $u(t)$ is

$$u_{t_1=.5}(t) = -(1\ 3) \begin{pmatrix} e^t & 0 \\ 0 & e^{2t} \end{pmatrix} W^{-1}(0, .5) \begin{pmatrix} 10 \\ -2 \end{pmatrix}$$

$$= -660.1278 e^t + 480.0624 e^{2t}.$$

The state vector $\vec{x}(t)$ was found numerically using MATLAB. The plot of $\vec{x}(t)$ for $t_1 = 1$ is given in Figure 7.2, and for $t_1 = .5$ in Figure 7.3. The controls $u_{t_1=1}$ and $u_{t_1=.5}$ are plotted on the same graph for comparison; note that the amplitude of $u_{t_1=.5}$ is greater that that of $u_{t_1=1}$ since it must drive the system to the desired final state in a shorter period of time (see Figure 7.4).

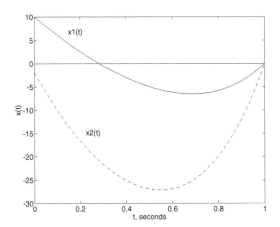

FIGURE 7.2: State vector $x(t)$ for $t_1 = 1$ second.

For ease of referencing, we will denote the time-invariant system (7.8) simply by the pair (A, B) when discussing controllability.

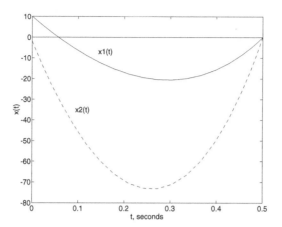

FIGURE 7.3:   State vector $x(t)$ for $t_1 = .5$ second.

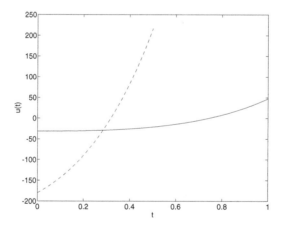

FIGURE 7.4:   Control $u(t)$ for $t_1 = 1$ second (solid line) and $t_1 = .5$ second (dashed line).

Consider next the example described by equations (7.1) and (7.2)

$$\begin{pmatrix} \dot{x}_1(t) \\ \dot{x}_2(t) \end{pmatrix} = \begin{pmatrix} -1 & 0 \\ 0 & 2 \end{pmatrix} \begin{pmatrix} x_1(t) \\ x_2(t) \end{pmatrix} + \begin{pmatrix} 1 \\ 0 \end{pmatrix} u(t),$$

$$y(t) = \begin{pmatrix} 0 & 1 \end{pmatrix} \begin{pmatrix} x_1(t) \\ x_2(t) \end{pmatrix}.$$

Simple calculations show that the controllability matrix is rank deficient (i.e., has less than full rank). We note, however, that the state variable x_2 is unaffected by the input u and hence the state space $(x_1, x_2)^T$ can be thought of as being decomposed into two subsystems, one being controllable (x_1) and the other uncontrollable (x_2). This is the well-known Kalman controllable canonical decomposition.

We will now outline a procedure for computing the similarity transformation matrix P to transform the time-invariant system (7.8) where $\rho(Q) = n_1$, with $n_1 < n$ into an equivalent system of the form

$$\begin{pmatrix} \dot{\tilde{x}}_1 \\ \dot{\tilde{x}}_2 \end{pmatrix} = \begin{pmatrix} A_1 & A_2 \\ 0 & A_3 \end{pmatrix} \begin{pmatrix} \tilde{x}_1 \\ \tilde{x}_2 \end{pmatrix} + \begin{pmatrix} B_1 \\ 0 \end{pmatrix} u(t) \tag{7.12}$$

where $\tilde{x}_1 \in \mathbb{R}^{n_1}$, $\tilde{x}_2 \in \mathbb{R}^{n-n_1}$, and (A_1, B_1) is controllable.

Let

$$\rho(Q) = \rho([B|AB|\dots|A^{n-1}B]) = n_1,$$

and

$$\tilde{P} = [P_1|P_2], \tag{7.13}$$

where P_1 is an $n \times n_1$ matrix whose columns form an orthogonal basis for the column space of Q and P_2 is an $n \times (n - n_1)$ matrix whose columns, in conjunction with those of P_1, form an orthogonal basis for $\mathbb{R}^n$.

Consider the state variable transformation

$$\tilde{P}\tilde{x}(t) = x(t)$$

which yields the equivalent system

$$\dot{\tilde{x}} = \tilde{P}^T A \tilde{P} \tilde{x} + \tilde{P}^T B u(t).$$

Partitioning this system according to (7.13) we obtain

$$\begin{pmatrix} \dot{\tilde{x}}_1 \\ \dot{\tilde{x}}_2 \end{pmatrix} = \begin{pmatrix} P_1{}^T A P_1 & P_1{}^T A P_2 \\ P_2{}^T A P_1 & P_2{}^T A P_2 \end{pmatrix} \begin{pmatrix} \tilde{x}_1 \\ \tilde{x}_2 \end{pmatrix} + \begin{pmatrix} P_1{}^T B \\ P_2{}^T B \end{pmatrix} u(t). \tag{7.14}$$

We next note that

(i) $P_1{}^T B = B_1$ for some appropriate $n_1 \times m$ matrix B_1. From the QR-decomposition, the controllability matrix Q is

$$Q = [B|AB|\dots|A^{n-1}B]$$
$$= \tilde{P}R$$
$$= [P_1|P_2] \begin{bmatrix} R_{11} & R_{12} & \dots & R_{1n} \\ R_{21} & R_{22} & \dots & R_{2n} \end{bmatrix},$$

where R has been partitioned such that R_{1i} is an $n_1 \times m$ matrix and R_{2i} is an $(n - n_1) \times m$ matrix for $i = 1, \dots, n$. This implies that $B =$

$P_1 R_{11} + P_2 R_{21}$. Therefore,

$$P_1^T B = P_1^T [P_1 R_{11} + P_2 R_{21}]$$
$$= P_1^T P_1 R_{11} + P_1^T P_2 R_{21}$$
$$= R_{11}$$

due to the orthogonality of the columns of P_1 and P_2.

(ii) $P_2^T B = P_2^T P_1 R_{11} = 0$. This again follows from the orthogonality of the columns of P_1 and P_2.

(iii) $P_2^T A P_1 = 0$.

Remark 7.2.4

(a) From (7.12), it is easy to see that the state variable $\tilde{x}_2(t)$ is completely unaffected by the input $u(t)$. Thus the state space has been divided into two parts, one being controllable and the other uncontrollable. This explains the term "controllable" used in Kalman controllable canonical transformation.

(b) A straightforward method of finding the orthogonal basis vectors in P_1 and P_2 is to calculate the QR decomposition of the controllability matrix (see Example 7.2.2). In addition, this method is a reliable technique to determine the rank of the controllability matrix. That is, one or more zeros on the diagonal in R implies that R and, consequently, the controllability matrix, do not have full rank.

Example 7.2.2 Consider the time-invariant system (7.8) where

$$A = \begin{pmatrix} 3 & 6 & 4 \\ 9 & 6 & 10 \\ -7 & -7 & -9 \end{pmatrix},$$

and

$$B = \begin{pmatrix} -0.666667 & 0.333333 \\ 0.333333 & -0.666667 \\ 0.333333 & 0.333333 \end{pmatrix}.$$

Let us first compute the controllability matrix, which in this example will be denoted by P (in order to avoid confusion with the notation Q used in the QR decomposition).

$$P = (B|AB|A^2 B)$$
$$= \begin{pmatrix} -0.6667 & 0.3333 & 1.3333 & -1.6667 & -2.6667 & 6.3333 \\ 0.3333 & -0.6667 & -0.6667 & 2.3333 & 1.3333 & -7.6667 \\ 0.3333 & 0.3333 & -0.6667 & -0.6667 & 1.3334 & 1.3334 \end{pmatrix}.$$

The rank of P, computed from MATLAB, is given as $\rho(P) = 3$. The QR decomposition of P is $P = QR$, where

$$Q = \begin{pmatrix} -0.8165 & 0.0000 & 0.5773 \\ 0.4082 & -0.7071 & 0.5773 \\ 0.4082 & 0.7071 & 0.5774 \end{pmatrix},$$

$$R = \begin{pmatrix} 0.8165 & -0.4082 & -1.6330 & 2.0412 & 3.2660 & -7.7567 \\ 0 & 0.7071 & 0.0000 & -2.1213 & 0.0000 & 6.3640 \\ 0 & 0 & 0.0000 & 0.0000 & 0.0000 & 0.0000 \end{pmatrix}.$$

Note, from the QR decomposition, $\rho(R) = 2$ and hence $\rho(P) = 2$, which is in conflict with MATLAB's computation of rank of (P). The reason for this discrepancy is that the MATLAB routine to calculate the rank of a matrix is very sensitive to round-off error. If we were to compute the rank of P using

$$B = \begin{pmatrix} -2/3 & 1/3 \\ 1/3 & -2/3 \\ 1/3 & 1/3 \end{pmatrix}$$

in MATLAB, the rank would be calculated correctly as 2. The MATLAB routine to compute the QR decomposition of a matrix A is $\mathtt{qr}(A)$.

We form the matrix $\tilde{P}$ as in (7.13) by letting P_1 consist of the first two columns of Q and P_2 be the third column of Q. Then $P = [P_1|P_2]$ has the desired properties, and

$$P_1 = \begin{pmatrix} -0.8165 & 0.0000 \\ 0.4082 & -0.7071 \\ 0.4082 & 0.7071 \end{pmatrix},$$

$$P_2 = \begin{pmatrix} 0.5773 \\ 0.5773 \\ 0.5774 \end{pmatrix}.$$

Hence, using equation (7.14) to find A_1, A_2, A_3, and B_1, we have

$$\begin{pmatrix} \dot{\tilde{x}}_1 \\ \dot{\tilde{x}}_2 \end{pmatrix} = \left(\begin{array}{c|c} A_1 & A_2 \\ \hline 0 & A_3 \end{array} \right) \begin{pmatrix} \tilde{x}_1 \\ \tilde{x}_2 \end{pmatrix} + \begin{pmatrix} B_1 \\ 0 \end{pmatrix} u(t)$$

$$= \left(\begin{array}{cc|c} -2.0000 & 1.7321 & -5.6569 \\ 0.0000 & -3.0000 & -19.5959 \\ \hline 0.0000 & 0.0000 & 5.0000 \end{array} \right) \begin{pmatrix} \tilde{x}_1 \\ \tilde{x}_2 \end{pmatrix} + \left(\begin{array}{cc} 0.8165 & -0.4082 \\ 0.0000 & 0.7071 \\ \hline 0.0000 & 0.0000 \end{array} \right) u(t),$$

where $\tilde{x}_1 \in \mathbb{R}^2$, $\tilde{x}_2 \in \mathbb{R}$, and (A_1, B_1) is controllable. To verify controllability, observe that the controllability matrix $\tilde{Q}$ of the subsystem (A_1, B_1) has full rank:

$$\rho(\tilde{Q}) = \rho([B_1|A_1 B_1]) = 2.$$

7.2.2 Observability

Closely linked to the idea of controllability is the concept of observability. In fact, these two concepts are dual. Loosely speaking, controllability studies the possibility of steering the state from the input; observability studies the possibility of determining the state of a system from the output. If a dynamical equation is controllable, all the modes of the equation can be excited from the input; if a dynamical equation is observable, all the modes of the equation can be observed from the output.

7.2.2.1 Time-Varying Case

Consider the n-dimensional linear state and output equations

$$\dot{\vec{x}}(t) = A(t)\vec{x}(t) + B(t)\vec{u}(t)$$
$$\vec{y}(t) = C(t)\vec{x}(t), \tag{7.15}$$

where $\vec{x}(\cdot) \in \mathbb{R}^n$, $\vec{u}(\cdot) \in \mathbb{R}^m$, $\vec{y}(\cdot) \in \mathbb{R}^p$, and $A(\cdot), B(\cdot)$, and $C(\cdot)$ are matrices of appropriate dimensions whose elements are continuous functions on $(-\infty, \infty)$.

Definition 7.2.2 *The dynamical system (7.15) is said to be (completely) observable at t_0 if there exists a finite time $t_1 > t_0$ such that for any initial state $\vec{x}(t_0) = \vec{x_0}$, the knowledge of $\vec{u}(t)$ and $\vec{y}(t)$ for $t \in [t_0, t_1]$ suffices to determine the state $\vec{x_0}$ uniquely.*

Example 7.2.3 Consider the system described by

$$\begin{pmatrix} \dot{x}_1 \\ \dot{x}_2 \end{pmatrix} = \begin{pmatrix} a_1 & 0 \\ 0 & a_2 \end{pmatrix} \begin{pmatrix} x_1 \\ x_2 \end{pmatrix} + \begin{pmatrix} b_1 \\ b_2 \end{pmatrix} u(t),$$

$$y(t) = x_2(t).$$

Because the system is decoupled and $y(t) = x_2(t)$, the state $x_1(t_0)$ cannot be determined by measuring $x_2(t)$ $(= y(t))$. Hence, the system is not observable.

We now determine conditions that can guarantee observability of (7.15). To this end, by the variation of constants formula we have

$$\vec{y}(t) = C(t)\left[\Phi(t, t_0)\vec{x_0} + \int_{t_0}^t \Phi(t, s)B(s)\vec{u}(s)ds\right].$$

In the study of observability, $\vec{y}(t)$ and $\vec{u}(t)$ are known functions (or measurements). Hence, the above equation can be rewritten as

$$C(t)\Phi(t, t_0)\vec{x_0} = \tilde{y}(t), \tag{7.16}$$

where $\tilde{y}(t)$ is a known function on $[t_0, t_1]$ and is given by

$$\tilde{y}(t) = \vec{y}(t) - C(t)\int_{t_0}^t \Phi(t, s)B(s)\vec{u}(s)ds.$$

Question: Can we determine $\vec{x_0}$ from (7.16)?

Multiply both sides of (7.16) by $\Phi^T(t, t_0)C^T(t)$ and integrate from t_0 to t_1. This yields

$$\left[\int_{t_0}^{t_1} \Phi^T(t, t_0)C^T(t)C(t)\Phi(t, t_0)dt\right]\vec{x_0} = \int_{t_0}^{t_1} \Phi^T(t, t_0)C^T(t)\tilde{y}(t)dt.$$

Therefore, if the constant matrix

$$V(t_0, t_1) = \int_{t_0}^{t_1} \Phi^T(t, t_0)C^T(t)C(t)\Phi(t, t_0)dt$$

is nonsingular (or, equivalently [6], all columns of $C(t)\Phi(t, t_0)$ are linearly independent on $[t_0, t_1]$) then we can determine $\vec{x_0}$ uniquely. In fact, the linear independence of the columns of $C(t)\Phi(t, t_0)$ is also a necessary condition for observability. To see this, we assume that the system (7.15) is observable at time t_0 but there exists no time $t_1 > t_0$ such that the columns of $C(\cdot)\Phi(\cdot, t_0)$ are linearly independent on $[t_0, t_1]$. Hence, the equation

$$C(t)\Phi(t, t_0)\vec{\alpha} = 0$$

has a nonzero $n \times 1$ constant vector solution for all $t > t_0$. Consider

$$\tilde{y}(t) = C(t)\Phi(t, t_0)\vec{x}(t_0)$$

for $t > t_0$. Since we assume (7.15) is observable, by taking $\vec{x}(t_0) = \vec{\alpha}$, we have

$$\vec{y}(t) = C(t)\Phi(t, t_0)\vec{\alpha} = 0$$

for all $t > t_0$, which implies that $\vec{\alpha}$ cannot be detected at the output. This contradicts the assumption of observability.

Remark 7.2.5 The above result shows that observability depends only on the matrices $C(\cdot)$ and $\Phi(\cdot, t_0)$, or equivalently, only on $C(\cdot)$ and $A(\cdot)$. Hence, in studying observability, it is convenient to assume that $u(t) \equiv 0$ and to refer to (7.15) by the pair (A, C).

The results on controllability and observability suggest that for controllability we study the rows of $\Phi(t_0, \cdot)B(\cdot)$ and for observability one considers the columns of $C(\cdot)\Phi(\cdot, t_0)$. These two concepts are in fact related by the well known Kalman Duality Theorem [6]. That is, consider the system

$$\begin{aligned}\dot{\vec{x}}(t) &= A(t)\vec{x}(t) + B(t)\vec{u}(t) \\ \vec{y}(t) &= C(t)\vec{x}(t)\end{aligned} \tag{7.17}$$

and the dual system

$$\begin{aligned}\dot{\vec{z}}(t) &= -A^T(t)\vec{z}(t) + C^T(t)\vec{v}(t) \\ \vec{w}(t) &= B^T(t)\vec{z}(t).\end{aligned} \tag{7.18}$$

System (7.17) is controllable (observable) at time t_0 if and only if the dual system (7.18) is observable (controllable) at time t_0 [6].

The Kalman duality result is very useful. It allows us to deduce from a controllability result the corresponding one on observability, and vice versa. For example, assume that the system matrix $A(\cdot)$ and the output matrix $C(\cdot)$ are $(n-1)$ times continuously differentiable. Then (7.15) is observable at time t_0 if there exists a finite time $t_1 > t_0$ such that

$$\rho \begin{pmatrix} N_0(t_1) \\ N_1(t_1) \\ \vdots \\ N_{n-1}(t_1) \end{pmatrix} = n,$$

where

$$N_0(t) = C(t)$$

$$N_{k+1}(t) = N_k(t)A(t) + \frac{d}{dt}N_k(t),$$

for $k = 0, 1, \ldots, n$.

7.2.2.2 Time-Invariant Case

Consider the linear time-invariant dynamical equation

$$\begin{aligned} \dot{\vec{x}}(t) &= A\vec{x}(t) \\ \vec{y}(t) &= C\vec{x}(t). \end{aligned} \tag{7.19}$$

As in the controllability case, if (A, C) is observable then it is observable at every $t_0 \geq 0$, and the determination of the initial state can be achieved in any finite time interval. Hence we drop the reference to t_0 and t_1 when we discuss observability of linear time-invariant systems.

From the Kalman duality theorem, the following *equivalent* statements can be easily obtained:

(i) The system (7.19) is observable;

(ii) The n columns of Ce^{At} are linearly independent on $[0, \infty)$;

(iii) The observability Grammian matrix

$$V(0, t) = \int_0^t e^{A^T s} C^T C e^{As} ds$$

is nonsingular for any $t > 0$. Furthermore, the initial state $\vec{x}(0) = \vec{x_0}$ can be determined from

$$\vec{x_0} = V^{-1}(0, t_1) \int_0^{t_1} e^{A^T t} C^T y(t)dt;$$

(iv) $\rho(\hat{V}) = n$, where $\hat{V}$ is the $pn \times n$ observability matrix

$$\hat{V} = \begin{pmatrix} C \\ CA \\ \vdots \\ CA^{n-1} \end{pmatrix}. \tag{7.20}$$

Example 7.2.4 Consider a spring-mass system with no damping, described by

$$\begin{pmatrix} \dot{x}_1 \\ \dot{x}_2 \end{pmatrix} = \begin{pmatrix} 0 & 1 \\ -1 & 0 \end{pmatrix} \begin{pmatrix} x_1 \\ x_2 \end{pmatrix}$$
$$y(t) = x_1(t).$$

Note that we observe the displacement. The observability matrix is

$$\begin{pmatrix} C \\ CA \end{pmatrix} = \begin{pmatrix} 1 & 0 \\ 0 & 1 \end{pmatrix},$$

which has rank 2 and, consequently, the system is observable. Let $t_0 = -\pi$, $t_1 = 0$, and suppose that we measure $y(t)$ to be

$$y(t) = \frac{1}{2} \cos t + \frac{1}{2} \sin t$$

on $[-\pi, 0]$. We seek to find $x_1(-\pi)$ and $x_2(-\pi)$.

We first note that $x_1(t) = y(t)$ and hence $x_1(-\pi) = -\frac{1}{2}$. Since $x_2 = \dot{x}_1 = \dot{y}$, then $x_2(-\pi) = -\frac{1}{2}$. Here, to find x_2 we need to differentiate y which is an unstable process (in practice, $y(t)$ has errors which are magnified by differentiation). We first compute

$$\Phi(t, -\pi) = \begin{pmatrix} -\cos t & -\sin t \\ \sin t & -\cos t \end{pmatrix},$$

so

$$C\Phi(t, -\pi) = \begin{pmatrix} -\cos t & -\sin t \end{pmatrix},$$

and

$$\begin{aligned} V(-\pi, 0) &= \int_{-\pi}^{0} \begin{pmatrix} -\cos t \\ -\sin t \end{pmatrix} \begin{pmatrix} -\cos t & -\sin t \end{pmatrix} dt \\ &= \frac{\pi}{2} \begin{pmatrix} 1 & 0 \\ 0 & 1 \end{pmatrix}. \end{aligned}$$

Therefore,

$$\begin{aligned} \begin{pmatrix} x_1(-\pi) \\ x_2(-\pi) \end{pmatrix} &= V^{-1}(-\pi, 0) \int_{-\pi}^{0} e^{A^T t} C^T y(t) dt \\ &= \frac{2}{\pi} \begin{pmatrix} 1 & 0 \\ 0 & 1 \end{pmatrix} \int_{-\pi}^{0} \begin{pmatrix} -\cos t \\ -\sin t \end{pmatrix} \left(\frac{1}{2} \cos t + \frac{1}{2} \sin t \right) dt \\ &= \begin{pmatrix} -\frac{1}{2} \\ -\frac{1}{2} \end{pmatrix} \end{aligned}$$

as we had computed earlier.

As in the controllability case, if the $\rho(\hat{V}) = n_1 < n$, where $\hat{V}$ is the observability matrix given by (7.20), then the state space can be divided into two subsystems — one observable, and one unobservable. This is analogously called the Kalman observable canonical decomposition. More precisely, consider the time-invariant system

$$\dot{\vec{x}}(t) = A\vec{x}(t) + B\vec{u}(t)$$
$$\vec{y}(t) = C\vec{x}(t),$$

where $\rho(\hat{V}) = n_1 < n$ and $\hat{V}$ is the observability matrix. Then there exists an equivalent system of the form

$$\begin{pmatrix} \dot{\tilde{x}}_1 \\ \dot{\tilde{x}}_2 \end{pmatrix} = \begin{pmatrix} A_1 & 0 \\ A_2 & A_3 \end{pmatrix} \begin{pmatrix} \tilde{x}_1 \\ \tilde{x}_2 \end{pmatrix} + \begin{pmatrix} B_1 \\ B_2 \end{pmatrix} u$$
$$\vec{y} = C_1 \tilde{x}_1,$$

where $\tilde{x}_1 \in \mathbb{R}_1^n$, $\tilde{x}_2 \in \mathbb{R}^{n-n_1}$ and (A_1, C_1) is observable.

Remark 7.2.6 Since $\rho(\hat{V}) = \rho(\hat{V}^T)$, the same QR decomposition applied to $\hat{V}^T$ can be used to find the required transformation matrix.

Example 7.2.5 Consider a linear time-invariant system where

$$A = \begin{pmatrix} 3 & 6 & 4 \\ 9 & 6 & 10 \\ -7 & -7 & -9 \end{pmatrix}, \qquad C = \begin{pmatrix} 1 & 2 & 3 \\ 3 & 3 & 6 \end{pmatrix}.$$

The transpose of the observability matrix is

$$\hat{V}^T = (C^T | A^T C^T | A^{2^T} C^T)$$
$$= \begin{pmatrix} 1 & 3 & 0 & -6 & -6 & 12 \\ 2 & 3 & -3 & -6 & 3 & 12 \\ 3 & 6 & -3 & -12 & -3 & 24 \end{pmatrix},$$

which has rank 2. Hence the state space for the system in $\mathbb{R}^3$ can be decomposed into two subsystems, where one state $\tilde{x}_1 \in \mathbb{R}^2$ is observable, and one state $\tilde{x}_2 \in \mathbb{R}^1$ is unobservable. The QR decomposition of $\hat{V}^T$ is $\hat{V}^T = QR$, where

$$Q = \begin{pmatrix} -0.2673 & 0.7715 & -0.5774 \\ -0.5345 & -0.6172 & -0.5774 \\ -0.8018 & 0.1543 & 0.5774 \end{pmatrix}$$

$$R = \begin{pmatrix} -3.7417 & -7.2161 & 4.0089 & 14.4321 & 2.4054 & -28.8642 \\ 0 & 1.3887 & 1.3887 & -2.7775 & -6.9437 & 5.5549 \\ 0 & 0 & 0.0000 & 0 & 0.0000 & 0 \end{pmatrix}.$$

We form the matrix $\tilde{P} = [P_1|P_2]$ as in Example 7.2.2, where P_1 consists of the first two columns of Q, and P_2 the third column of Q. Then the equivalent system under the state variable transformation $\tilde{P}\tilde{x}(t) = x(t)$ is

$$\begin{pmatrix} \dot{\tilde{x}}_1 \\ \dot{\tilde{x}}_2 \end{pmatrix} = \begin{pmatrix} P_1^T A P_1 & P_1^T A P_2 \\ P_2^T A P_1 & P_2^T A P_2 \end{pmatrix} \begin{pmatrix} \tilde{x}_1 \\ \tilde{x}_2 \end{pmatrix} + \begin{pmatrix} P_1^T B \\ P_2^T B \end{pmatrix} u(t)$$

$$= \begin{pmatrix} -1.0714 & -0.3712 & 0.0000 \\ 4.8250 & -3.9286 & 0.0000 \\ 19.4422 & -3.7417 & 5.0000 \end{pmatrix} \begin{pmatrix} \tilde{x}_1 \\ \tilde{x}_2 \end{pmatrix} + \begin{pmatrix} -3.7417 & -7.2161 \\ 0.0000 & 1.3887 \\ 0 & 0.0000 \end{pmatrix} u(t)$$

and

$$\vec{y} = [C_1|C_2] \begin{pmatrix} \tilde{x}_1 \\ \tilde{x}_2 \end{pmatrix}$$

$$= \begin{pmatrix} -3.7417 & 0.0000 & 0 \\ -7.2161 & 1.3887 & 0 \end{pmatrix} \begin{pmatrix} \tilde{x}_1 \\ \tilde{x}_2 \end{pmatrix}$$

$$= C_1 \tilde{x}_1,$$

where $C_1 = [CP_1]$. To verify observability, note that the observability matrix V_1 of the subsystem (A_1, C_1) has full rank:

$$\rho(V_1) = \rho(V_1^T) = \rho([C_1^T|A_1^T C_1^T]) = 2.$$

7.3 Design of State Feedback Control Systems and State Estimators

We begin this section by reviewing the concept of stability. Consider the linear time-invariant controlled system $\dot{x}(t) = Ax(t) + Bu(t)$ with initial condition $x(t_0) = x_0$. For $u(t) = 0$, the solution is given by

$$x(t) = \Phi(t; t_0, x_0)$$
$$= \Phi(t, t_0)x_0.$$

Definition 7.3.1 *A state x_e of a dynamical equation is said to be an equilibrium state at t_0 if*

$$x_e = \Phi(t; t_0, x_e)$$

for all $t \geq t_0$.

Therefore, if a trajectory reaches an equilibrium state and no input is applied, the trajectory will stay at the equilibrium state forever; that is, $\dot{x}_e(t) = 0$ for all $t \geq t_0$. To find x_e, set the right side of the differential equation equal to zero. For example, to find x_e for $\dot{x}(t) = A(t)x(t)$ we solve

$$A(t)x(t) = 0.$$

Hence, $x(t) = 0$ is always an equilibrium state of $\dot{x}(t) = A(t)x(t)$.

Definition 7.3.2

(a) *An equilibrium state x_e is said to be* stable in the sense of Lyapunov
    *(i.s.L.) at t_0 if and only if for every $\varepsilon > 0$ there exists a $\delta(\varepsilon, t_0) > 0$
    (which depends on ε and t_0) such that if $\|x_0 - x_e\| < \delta(\varepsilon, t_0)$ then
    $\|\Phi(t; t_0, x_0) - x_e\| < \varepsilon$ for all $t \geq t_0$.*

(b) *If δ depends only on ε but not on t_0, then we say that x_e is* uniformly
    stable i.s.L.

Basically, x_e is stable i.s.L. if the response due to any initial state that is
sufficiently near x_e does not move far away from x_e.

Remark 7.3.1 The state x_e is uniformly stable i.s.L. (u.s.i.s.L.) implies it is
stable i.s.L. (s.i.s.L.). The converse may not be true.

Example 7.3.1 Consider a pendulum as depicted in Figure 7.5. Applying
Newton's second law of motion yields

$$u(t) \cos \theta - mg \sin \theta = ml\ddot{\theta}.$$

Let $x_1 = \theta$, and $x_2 = \dot{\theta}$. Then

$$\frac{d}{dt} \begin{pmatrix} x_1 \\ x_2 \end{pmatrix} = \begin{pmatrix} x_2 \\ -\frac{g}{l} \sin x_1 + \frac{\cos x_1}{ml} u \end{pmatrix}.$$

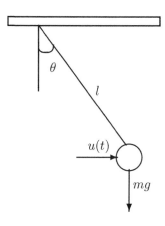

FIGURE 7.5: A pendulum.

For the equilibrium state, we take $u = 0$, and set $\dfrac{d}{dt}\begin{pmatrix} x_1 \\ x_2 \end{pmatrix} = \vec{0}$: we find

$$\begin{pmatrix} x_2 \\ \frac{g}{l}\sin x_1 \end{pmatrix} = \begin{pmatrix} 0 \\ 0 \end{pmatrix} \text{ which implies } x = \begin{pmatrix} k\pi \\ 0 \end{pmatrix}, k = 0, \pm 1, \pm 2, \ldots$$

Note the equilibrium states satisfy

- $x_e = \begin{pmatrix} k\pi \\ 0 \end{pmatrix}, \quad k = 0, \pm 2, \ldots,$ are u.s.i.s.L.

- $x_e = \begin{pmatrix} k\pi \\ 0 \end{pmatrix}, \quad k = \pm 1, \pm 3, \ldots,$ are not s.i.s.L. (Why?)

Definition 7.3.3

(a) *An equilibrium state x_e is said to be* asymptotically stable *at t_0 if it is stable i.s.L. at t_0 and if every motion starting sufficiently near x_e converges to x_e as $t \to \infty$; that is, there exists a $\gamma > 0$ such that if $\|x(t_1) - x_e\| \le \gamma$, then for any $\bar{\varepsilon} > 0$ there exists $T(\gamma, \bar{\varepsilon}, t_1) > 0$ (that depends on $\gamma, \bar{\varepsilon}, t_1$) such that*

$$\|\Phi(t; t_1, x(t_1)) - x_e\| \le \bar{\varepsilon}$$

for all $t \ge t_1 + T(\gamma, \bar{\varepsilon}, t_1)$.

(b) *If an equilibrium state x_e is u.s.i.s.L. and T can be chosen independent of t_1 in the definition of asymptotic stability, then we say that x_e is* uniformly asymptotically stable *over $[t_0, \infty)$.*

Remark 7.3.2 For the linear time-invariant system, it can be shown that [6]:

(a) Every equilibrium state of $\dot{x}(t) = Ax(t)$ is s.i.s.L. if and only if

    – all eigenvalues of A have nonpositive real parts (negative or zero).

    – for any eigenvalue on the imaginary axis $(\mathrm{Re}(\lambda) = 0)$ with multiplicity m there correspond exactly m eigenvectors of A.

(b) The zero state of $\dot{x}(t) = Ax(t)$ is asymptotically stable if and only if all the eigenvalues of A have negative real parts.

Let us consider a linear time-invariant control system

$$\dot{\vec{x}}(t) = A\vec{x}(t) + B\vec{u}(t), \tag{7.21}$$

where $x(\cdot) \in \mathbb{R}^n$, $u(\cdot) \in \mathbb{R}^m$ and A, B are matrices of appropriate dimensions. The system (7.21) may often arise as the linearization of some nonlinear system about an equilibrium point or about the original system dynamics of interest. Now assume that the homogeneous system $(u \equiv 0)$

$$\dot{\vec{x}}(t) = A\vec{x}(t) \tag{7.22}$$

is not asymptotically stable. In control theory, the aim is to compel or control a system to behave in some desired fashion. Thus for system (7.21), an objective would be to use the control $\vec{u}(\cdot)$ so that the system becomes asymptotically stable. The traditionally favored means of accomplishing this task is to use a feedback relation

$$\vec{u} = K\vec{x},$$

where K is an $m \times n$ matrix. The problem is thus to find the *gain* or *feedback* matrix K so that

$$\begin{aligned}
\dot{\vec{x}}(t) &= A\vec{x}(t) + BK\vec{x}(t) \\
&= (A + BK)\vec{x}(t)
\end{aligned} \tag{7.23}$$

is asymptotically stable (i.e., every eigenvalue of $A + BK$ has a negative real part). The system (7.23) is called a *closed-loop* or *feedback* control system. The main features of a feedback control system are represented in Figure 7.6.

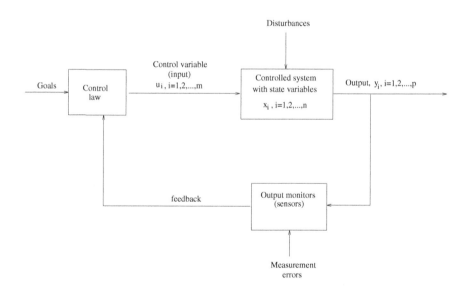

FIGURE 7.6:   A closed-loop or feedback control system.

Another class of control systems, called an *open-loop* control system, is represented in Figure 7.7.

In an open-loop system, the control $\vec{u}(\cdot)$ is computed based on the goals for the system and all available *a priori* knowledge about the system. The input $\vec{u}(\cdot)$ is in no way influenced by the output $\vec{y}(\cdot)$ and thus if unexpected disturbances act upon the system or there are changes in operating conditions,

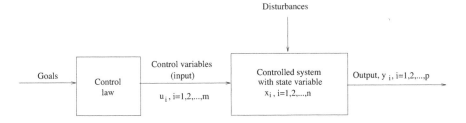

FIGURE 7.7: An open-loop control system.

the output $\vec{y}(\cdot)$ will not behave precisely as desired. On the other hand, in a closed-loop system there is a feedback of information concerning the outputs to the controller. Thus, a feedback system is better to adapt to changes in the system parameters or to unexpected disturbances. However, if the measurement errors are large, closed-loop control performance might be inferior to open-loop control.

7.3.1   Effect of State Feedback on System Properties

In this section we will discuss how state feedback affects system properties such as stability, controllability and observability.

7.3.1.1   Stability

Stability of a linear time-invariant system depends entirely on the location of the eigenvalues of the system matrix. A feedback control law yields closed-loop eigenvalues which differ from the open-loop eigenvalues. In addition, small time delays in state feedback can destabilize a system which is asymptotically stable in the absence of such delays. Such time delays might occur in computing feedback controls.

Example 7.3.2 Consider the linear, time-invariant system

$$\dot{\vec{x}}(t) = \begin{pmatrix} 2 & 1 \\ 0 & 1 \end{pmatrix} \vec{x}(t) + \begin{pmatrix} 0 \\ 1 \end{pmatrix} u(t)$$

that has open-loop eigenvalues $\lambda_1 = 1$ and $\lambda_2 = 2$ (unstable). Now consider a state feedback control law of the form

$$u = K\vec{x}$$

where $K = [-9 \ -5]$. The actual computation of K will be discussed later in this chapter. The closed-loop system is then

$$\dot{\vec{x}}(t) = \left[\begin{pmatrix} 2 & 1 \\ 0 & 1 \end{pmatrix} + \begin{pmatrix} 0 \\ 1 \end{pmatrix} \begin{pmatrix} -9 & -5 \end{pmatrix} \right] \vec{x}(t)$$

$$= \begin{pmatrix} 2 & 1 \\ -9 & -4 \end{pmatrix} \vec{x}(t),$$

which has eigenvalues $\tilde{\lambda}_1 = \tilde{\lambda}_2 = -1$ and thus is asymptotically (exponentially) stable. Let us now assume that there is a small time delay in the feedback loop of the form

$$u(t) = K\vec{x}(t - h),$$

where $h > 0$. The closed-loop system then becomes

$$\dot{\vec{x}}(t) = \begin{pmatrix} 2 & 1 \\ 0 & 1 \end{pmatrix} \vec{x}(t) + \begin{pmatrix} 0 & 0 \\ -9 & -5 \end{pmatrix} \vec{x}(t - h) \qquad (7.24)$$

which is a delay differential equation. Taking the Laplace transform of (7.24) and assuming that $\vec{x}(0) = \vec{0}$ we obtain

$$\left[sI - \begin{pmatrix} 2 & 1 \\ 0 & 1 \end{pmatrix} - \begin{pmatrix} 0 & 0 \\ -9 & -5 \end{pmatrix} e^{-hs} \right] X(s) = 0,$$

where $X(s) = \mathcal{L}\{x(t)\}$, the Laplace transform of $x(t)$.

The eigenvalues of the closed-loop system are given by the roots of

$$\det \Delta(\lambda) = 0, \qquad (7.25)$$

where $\Delta(\lambda) = \lambda I - \begin{pmatrix} 2 & 1 \\ 0 & 1 \end{pmatrix} - \begin{pmatrix} 0 & 0 \\ -9 & -5 \end{pmatrix} e^{-h\lambda}$ (see e.g., [10]). We use a method described in [14] to compute the roots of (7.25). For $h = 0.1$, all roots of (7.25) have negative real parts. However, when $h = 0.22$, one root of (7.25) has a positive real part,

$$\lambda = 0.03122 \pm 4.43793i.$$

This example illustrates that a sufficiently large time delay, when introduced into the feedback law, can destabilize a system which is exponentially stable in the absence of delays. This destabilization of state feedback is also observed for certain infinite-dimensional systems (see [7, 8]).

7.3.1.2 Controllability

Consider the linear, time-invariant system (7.21) and assume that $\vec{u}(t)$ is decomposed into two parts

$$\vec{u}(t) = \tilde{u}(t) + \vec{u}_r(t),$$

where $\tilde{u}(t) = K\vec{x}(t)$ and $\vec{u}_r(t)$ is the reference input. The closed-loop system has the form

$$\dot{\vec{x}}(t) = (A + BK)\vec{x}(t) + B\vec{u}_r(t).$$

We now pose the following question: If (A, B) is controllable, is $(A + BK, B)$ controllable? That is, is controllability invariant under state feedback?

We note that the controllability matrix of the closed-loop system can be written as

$$(B|(A+BK)B|\ldots|(A+BK)^{n-1}B) = (B|AB|\ldots|A^{n-1}B) \times$$

$$\underbrace{\begin{pmatrix} I & KB & KAB+(KB)^2 & \cdots & * \\ 0 & I & KB & \cdots & * \\ \vdots & & & \ddots & \ddots & KB \\ 0 & \cdots & & & 0 & I \end{pmatrix}}_{M}.$$

Since the matrix M on the right side of the equation is non-singular, we have

$$\rho([B|(A+BK)B|\ldots|(A+BK)^{n-1}B]) = \rho([B|AB|\ldots|A^{n-1}B]).$$

Hence, this establishes that the pair $(A + BK, B)$ is controllable if and only if the pair (A, B) is controllable.

7.3.1.3 Observability

Analogously, we now want to know if observability is invariant under state feedback. Let us consider a particular example.

Example 7.3.3 Consider the following system

$$\dot{\vec{x}}(t) = \begin{pmatrix} 1 & 1 \\ 0 & 1 \end{pmatrix} \vec{x}(t) + \begin{pmatrix} b_1 \\ b_2 \end{pmatrix} u(t)$$

$$y(t) = \begin{pmatrix} c_1 & c_2 \end{pmatrix} \vec{x}(t).$$

The observability matrix is

$$V = \begin{pmatrix} C \\ CA \end{pmatrix} = \begin{pmatrix} c_1 & c_2 \\ c_1 & c_1 + c_2 \end{pmatrix}$$

which has rank 2 if and only if $c_1 \neq 0$ (c_2 is arbitrary). Hence, in particular, the system

$$\dot{\vec{x}}(t) = \begin{pmatrix} 1 & 1 \\ 0 & 1 \end{pmatrix} \vec{x}(t) + \begin{pmatrix} b_1 \\ b_2 \end{pmatrix} u(t)$$

$$y(t) = \begin{pmatrix} 1 & 0 \end{pmatrix} \vec{x}(t)$$

is observable. Now consider the feedback law $u(t) = K\tilde{x}(t)$. Is the pair $(A + BK, C)$ observable? The new observability matrix is

$$\tilde{V} = \begin{pmatrix} C \\ C(A + BK) \end{pmatrix} = \begin{pmatrix} 1 & 0 \\ 1 + b_1 k_1 & 1 + b_1 k_2 \end{pmatrix}$$

which has rank 2 if and only if $1 + b_1 k_2 \neq 0$.

Example 7.3.3 is an illustration of the general result: observability is *not* invariant under state feedback.

7.4   Pole Placement (Relocation) Problem

Equation (7.23) indicates that the eigenvalues of the closed-loop system using state feedback $\vec{u} = K\vec{x}$, where K is a constant $m \times n$ feedback matrix, are the roots of

$$|\lambda I - A - BK| = 0. \tag{7.26}$$

We now consider the following problem: Given a set of desired eigenvalues $\{\lambda_1^D, \lambda_2^D, \ldots, \lambda_n^D\}$ in the complex plane, can we find the state variable feedback (gain) matrix K so that the poles of $A_C \equiv A + BK$ are the desired ones? That is, can we relocate the poles of the open-loop system, $\lambda_1, \lambda_2, \ldots, \lambda_n$ (the roots of $|\lambda I - A| = 0$), to the desired locations $\lambda_1^D, \lambda_2^D, \ldots, \lambda_n^D$ in the complex plane by state feedback. We note that since the elements of $A_C = A + BK$ are real, if any desired poles are complex, they must appear in (complex) conjugate pairs.

The answer to the above question is provided by the following result, which gives an important link between controllability and pole placement using state feedback. If the linear, time-invariant control system (7.21) is controllable, then by the state feedback $\vec{u} = K\vec{x}$, where K is an $m \times n$ constant, real matrix, the eigenvalues of $A_C = A + BK$ can be arbitrarily placed anywhere in the complex plane. The proof of this result can be found in [9] and [16]. In the following, we will show how to construct the appropriate feedback gain K by two different methods.

(i) Direct Method
The goal is to have the following equality

$$|\lambda I - (A + BK)| = \prod_{i=1}^{n} (\lambda - \lambda_i^D).$$

Expanding both sides of the above equation, we find

$$a_n^C + a_{n-1}^C \lambda + \ldots + a_1^C \lambda^{n-1} + \lambda^n = a_n^D + a_{n-1}^D \lambda + \ldots + a_1^D \lambda^{n-1} + \lambda^n. \tag{7.27}$$

The coefficients a_i^C, for $i = 1, 2, \ldots, n$, are functions of the elements k_{ij}, for $i = 1, 2, \ldots, m$ and $j = 1, 2, \ldots, n$, of the feedback gain matrix K. By equating the coefficients of λ in (7.27) we obtain n equations in mn unknowns. This method illustrates that feedback gain in pole placement is unique only when $m = 1$ (the single input case). Otherwise, many gains give the same desired locations for closed-loop eigenvalues. We will first restrict our attention to the single-input case.

Example 7.4.1 Consider the system (7.21) where

$$A = \begin{pmatrix} 2 & 1 \\ 0 & 1 \end{pmatrix}, \qquad B = \begin{pmatrix} 0 \\ 1 \end{pmatrix}.$$

It is a simple exercise to show that $\rho([B|AB]) = 2$, which implies that (A, B) is controllable. Find the state variable feedback gain $K = \begin{pmatrix} k_1 & k_2 \end{pmatrix}$ so that the eigenvalues can be relocated anywhere in the complex plane.

Let λ_1^D and λ_2^D be arbitrarily given in $\mathbb{C}$ (if they have nonzero imaginary parts, $\lambda_1^D = \overline{\lambda_2^D}$). The problem is to find $K = \begin{pmatrix} k_1, & k_2 \end{pmatrix}$ so that

$$|\lambda I - (A + BK)| = (\lambda - \lambda_1^D)(\lambda - \lambda_2^D).$$

Expanding both sides, we obtain the following equation:

$$\lambda^2 - \lambda(2 + 1 + k_2) + 2(1 + k_2) - k_1 = \lambda^2 - (\lambda_1^D + \lambda_2^D)\lambda + \lambda_1^D \lambda_2^D.$$

Simplifying and equating the coefficients in λ we obtain the following two equations in the two unknowns k_1 and k_2:

$$\lambda_1^D + \lambda_2^D = 3 + k_2$$
$$\lambda_1^D \lambda_2^D = 2 + 2k_2 - k_1$$

from which we obtain

$$k_1 = -\lambda_1^D \lambda_2^D + 2(1 + k_2)$$
$$k_2 = (\lambda_1^D + \lambda_2^D) - 3.$$

For example, if $\lambda_1^D = -1$ and $\lambda_2^D = -1$, then $k_1 = -9$ and $k_2 = -5$, so

$$A + BK = \begin{pmatrix} 2 & 1 \\ -9 & -4 \end{pmatrix}$$

with characteristic polynomial

$$|\lambda I - (A + BK)| = (\lambda - 2)(\lambda + 4) + 9$$
$$= (\lambda + 1)^2.$$

Hence, the feedback gain $K = \begin{pmatrix} -9 & -5 \end{pmatrix}$ gives the desired closed-loop eigenvalues.

It should be emphasized that some eigenvalue relocation may be achievable with state feedback even when (A, B) is not controllable.

Example 7.4.2 Consider the system (7.21) with

$$A = \begin{pmatrix} 1 & 1 \\ 0 & -1 \end{pmatrix}, \qquad B = \begin{pmatrix} 1 \\ 0 \end{pmatrix}.$$

Here, the controllability matrix $[B|AB]$ has rank 1, and therefore (A, B) is not controllable. Nevertheless, let us now consider the state feedback, $\vec{u} = K\vec{x}$. Equating the coefficients of λ in

$$|\lambda I - (A + BK)| = (\lambda - \lambda_1^D)(\lambda - \lambda_2^D)$$

we get

$$k_1 = \lambda_1^D + \lambda_2^D$$
$$-(1 + k_1) = \lambda_1^D \lambda_2^D.$$

Hence, If $\lambda_1^D = -2$ and $\lambda_2^D = -1$, $k_1 = -3$ and k_2 arbitrary we will obtain the desired closed-loop eigenvalues. That is, just because (A, B) is not controllable does not mean that we cannot achieve some desired pole-relocations. However, if $\lambda_1^D = -2$ and $\lambda_2^D = -3$, then

$$k_1 = \lambda_1^D + \lambda_2^D = -5$$

but

$$-(1 + k_1) = 4 \neq \lambda_1^D \lambda_2^D = 6.$$

That is, the closed-loop poles $\lambda_1^D = -2$ and $\lambda_2^D = -3$ cannot be achieved with any feedback gain $K = \begin{pmatrix} k_1, k_2 \end{pmatrix}$.

(ii) Use of Controllable Canonical Form
Recall that a scalar n^{th}-order differential equation

$$\frac{d^n y}{dt^n} + a_1 \frac{d^{n-1} y}{dt^{n-1}} + \ldots + a_{n-1} \frac{dy}{dt} + a_n y = u(t)$$

can be written equivalently as a first-order system of differential equations as follows:

$$\begin{pmatrix} \dot{x}_1 \\ \dot{x}_2 \\ \dot{x}_3 \\ \vdots \\ \dot{x}_n \end{pmatrix} = \begin{pmatrix} 0 & 1 & 0 & \cdots & 0 \\ 0 & 0 & 1 & & 0 \\ 0 & 0 & 0 & \ddots & \vdots \\ \vdots & & & \ddots & 1 \\ -a_n & -a_{n-1} & \cdots & & -a_1 \end{pmatrix} \begin{pmatrix} x_1 \\ x_2 \\ x_3 \\ \vdots \\ x_n \end{pmatrix} + \begin{pmatrix} 0 \\ 0 \\ 0 \\ \vdots \\ 1 \end{pmatrix} u(t), \qquad (7.28)$$

where $x_1 = y, x_2 = \dot{y}, \dots, x_n = y^{n-1}$. We will now establish that for linear, time-invariant, single-input control systems, if (A, B) is controllable, then there exists a coordinate transformation such that the new equivalent dynamical equation has the form (7.28) which is called the *controllable canonical form*. The word "controllable" is used because (7.28) is indeed controllable. More precisely, consider the linear control system with single input,

$$\dot{\vec{x}}(t) = A\vec{x}(t) + \vec{b}u(t) \tag{7.29}$$

and suppose that $(A, \vec{b})$ is controllable. Then there is a nonsingular coordinate transformation

$$\vec{z}(t) = P^{-1}\vec{x}(t)$$

such that the equivalent dynamical equation

$$\dot{\vec{z}}(t) = P^{-1}AP\vec{z}(t) + P^{-1}\vec{b}u(t) \tag{7.30}$$

has the form (7.28). To construct the transformation matrix P^{-1} we denote $\vec{p} = \begin{pmatrix} p_1 & p_2 & \cdots & p_n \end{pmatrix}^T \in \mathbb{R}^n$ and assume that the coordinate transformation P^{-1} has the form

$$P^{-1} = \begin{pmatrix} \vec{p}^T \\ \vec{p}^T A \\ \vdots \\ \vec{p}^T A^{n-1} \end{pmatrix}$$

and P is partitioned as

$$P = \begin{pmatrix} \vec{q}_1 & \vec{q}_2 & \cdots & \vec{q}_n \end{pmatrix},$$

where $\vec{q}_i \in \mathbb{R}^n$. Since $P^{-1}P = I$, we have

$$\begin{pmatrix} \vec{p}^T \\ \vec{p}^T A \\ \vdots \\ \vec{p}^T A^{n-1} \end{pmatrix} \begin{pmatrix} \vec{q}_1 & \vec{q}_2 & \cdots & \vec{q}_n \end{pmatrix} = \begin{pmatrix} 1 & 0 & \cdots & 0 \\ 0 & 1 & \ddots & \vdots \\ \vdots & \ddots & \ddots & 0 \\ 0 & 0 & \cdots & 1 \end{pmatrix}$$

or

$$\begin{pmatrix} \vec{p}^T \vec{q}_1 & \vec{p}^T \vec{q}_2 & \cdots & \vec{p}^T \vec{q}_n \\ \vec{p}^T A\vec{q}_1 & \vec{p}^T A\vec{q}_2 & \cdots & \vec{p}^T A\vec{q}_n \\ \vdots & & & \vdots \\ \vec{p}^T A^{n-1}\vec{q}_1 & \vec{p}^T A^{n-1}\vec{q}_2 & \cdots & \vec{p}^T A^{n-1}\vec{q}_n \end{pmatrix} = \begin{pmatrix} 1 & 0 & \cdots & 0 \\ 0 & 1 & \ddots & \vdots \\ \vdots & \ddots & \ddots & 0 \\ 0 & 0 & \cdots & 1 \end{pmatrix}. \tag{7.31}$$

Also, we have

$$P^{-1}AP = \begin{pmatrix} \vec{p}^T \\ \vec{p}^T A \\ \vdots \\ \vec{p}^T A^{n-1} \end{pmatrix} \begin{pmatrix} A\vec{q}_1 & A\vec{q}_2 & \cdots & A\vec{q}_n \end{pmatrix}$$

$$= \begin{pmatrix} \vec{p}^T A\vec{q}_1 & \vec{p}^T A\vec{q}_2 & \cdots & \vec{p}^T A\vec{q}_n \\ \vec{p}^T A^2\vec{q}_1 & \vec{p}^T A^2\vec{q}_2 & \cdots & \vec{p}^T A^2\vec{q}_n \\ \vdots & & & \vdots \\ \vec{p}^T A^n\vec{q}_1 & \vec{p}^T A^n\vec{q}_2 & \cdots & \vec{p}^T A^n\vec{q}_n \end{pmatrix}.$$

After comparing the above matrix with (7.31) we obtain

$$P^{-1}AP = \begin{pmatrix} 0 & 1 & 0 & & 0 \\ 0 & 0 & 1 & \ddots & \vdots \\ & & & \ddots & \\ 0 & 0 & \cdots & & 1 \\ \vec{p}^T A^n\vec{q}_1 & \vec{p}^T A^n\vec{q}_2 & \cdots & & \vec{p}^T A^n\vec{q}_n \end{pmatrix}.$$

It therefore remains to force the condition

$$P^{-1}\vec{b} = \begin{pmatrix} \vec{p}^T \vec{b} \\ \vec{p}^T A\vec{b} \\ \vdots \\ \vec{p}^T A^{n-1}\vec{b} \end{pmatrix} = \begin{pmatrix} 0 \\ 0 \\ \vdots \\ 1 \end{pmatrix}$$

or, equivalently,

$$\vec{p}^T \begin{pmatrix} \vec{b} & A\vec{b} & \cdots & A^{n-1}\vec{b} \end{pmatrix} = \begin{pmatrix} 0 & 0 & \cdots & 1 \end{pmatrix}.$$

This is a system of n equations in n unknowns $\vec{p} = \begin{pmatrix} p_1 & p_2 & \cdots & p_n \end{pmatrix}^T$ which has a unique solution since the matrix $\begin{pmatrix} \vec{b} & A\vec{b} & \cdots & A^{n-1}\vec{b} \end{pmatrix}$ has full rank or is nonsingular.

Example 7.4.3 Transform the following control system

$$\dot{\vec{x}}(t) = \begin{pmatrix} 13 & 35 \\ -6 & -16 \end{pmatrix} \vec{x}(t) + \begin{pmatrix} -2 \\ 1 \end{pmatrix} u(t)$$

into controllable canonical form.

We first note that $\rho([B|AB]) = 2$ and, consequently, there exists a unique coordinate transformation matrix P such that the equivalent dynamical equation in the new coordinate is in controllable canonical form. Let $\varphi = \begin{pmatrix} p_1 & p_2 \end{pmatrix}^T$ and assume that

$$P^{-1} = \begin{pmatrix} \varphi^T \\ \varphi^T A \end{pmatrix}.$$

The vector φ is the unique solution of

$$\begin{pmatrix} \varphi^T \vec{b} \\ \varphi^T A \vec{b} \end{pmatrix} = \begin{pmatrix} 0 \\ 1 \end{pmatrix}$$

or

$$-2p_1 + p_2 = 0$$
$$9p_1 - 4p_2 = 1,$$

which has the solution $p_1 = 1$ and $p_2 = 2$. Therefore,

$$P^{-1} = \begin{pmatrix} 1 & 2 \\ 1 & 3 \end{pmatrix}, \qquad \text{and } P = \begin{pmatrix} 3 & -2 \\ -1 & 1 \end{pmatrix}.$$

We check that

$$P^{-1}AP = \begin{pmatrix} 0 & 1 \\ -2 & -3 \end{pmatrix}, \qquad \text{and} \qquad P^{-1}\vec{b} = \begin{pmatrix} 0 \\ 1 \end{pmatrix}$$

which is of controllable canonical form.

We note that the elements a_i, for $i = 1, 2, \ldots, n$, in the last row of $P^{-1}AP$ in (7.28) are also the coefficients in the characteristic polynomial of A; that is,

$$|\lambda I - A| = \lambda^n + a_1\lambda^{n-1} + a_2\lambda^{n-2} + \ldots + a_{n-1}\lambda + a_n.$$

The controllability matrices of (7.29) and (7.30) are, respectively, given by

$$Q = [\vec{b}|A\vec{b}|\cdots|A^{n-1}\vec{b}]$$

$$\begin{aligned} \tilde{Q} &= [P^{-1}\vec{b}|P^{-1}APP^{-1}\vec{b}|\cdots|P^{-1}A^{n-1}PP^{-1}\vec{b}] \\ &= P^{-1}[\vec{b}|A\vec{b}|\cdots|A^{n-1}\vec{b}] \\ &= P^{-1}Q. \end{aligned} \tag{7.32}$$

The above equation gives a link between the coordinate transformation and the controllability matrices. Hence, another way to compute P^{-1} is by

$$P^{-1} = \tilde{Q}Q^{-1},$$

where Q is nonsingular by the controllability assumption of (7.29).

Let the equivalent dynamical equation (7.30) be of the controllable canonical form (7.28) and let

$$\vec{u}(t) = K\vec{z}(t),$$

where $K = \begin{pmatrix} k_n & k_{n-1} & \cdots & k_1 \end{pmatrix}$. The closed-loop system is then given by

$$\dot{\vec{z}}(t) = (P^{-1}AP + P^{-1}\vec{b}K)\vec{z}(t),$$

where

$$A_C = P^{-1}AP + P^{-1}\vec{b}K$$

$$= \begin{pmatrix} 0 & 1 & 0 \cdots & & 0 \\ 0 & 0 & 1 & & \\ & & & \ddots & \vdots \\ & & & & 1 \\ -\tilde{a}_n & -\tilde{a}_{n-1} & \cdots & & -\tilde{a}_1 \end{pmatrix}$$

and

$$\tilde{a}_i = a_i - k_i.$$

The characteristic polynomial of A_C has the form

$$|\lambda I - A_C| = \lambda^n + \tilde{a}_1 \lambda^{n-1} + \cdots + \tilde{a}_{n-1}\lambda + \tilde{a}_n.$$

By equating the coefficients $\tilde{a}_i$ to those of $\prod_{i=1}^n (\lambda - \lambda_i^D)$ we can solve for the elements k_i of the feedback gain matrix K. We summarize the procedure of computing K in the following algorithm.

Algorithm

1. Find the characteristic polynomial of A:

$$|\lambda I - A| = \lambda^n + a_1 \lambda^{n-1} + \cdots + a_{n-1}\lambda + a_n.$$

2. Compute

$$\prod_{i=1}^n (\lambda - \lambda_i^D) = \lambda^n + a_1^D \lambda^{n-1} + \cdots + a_{n-1}^D \lambda + a_n^D.$$

3. Compute

$$k_i = a_i - a_i^D,$$

for $i = 1, 2, \ldots, n$.

4. Compute the coordinate transformation P^{-1} by (7.32).

5. The feedback law for the system in canonical form (7.30) is

$$\vec{u}(t) = K\vec{z}(t).$$

6. The feedback law for the original system (7.29) is

$$\vec{u}(t) = KP^{-1}\vec{x}(t).$$

Example 7.4.4 Consider the previous example where

$$\dot{\vec{x}}(t) = \begin{pmatrix} 13 & 35 \\ -6 & -16 \end{pmatrix} \vec{x}(t) + \begin{pmatrix} -2 \\ 1 \end{pmatrix} u(t)$$

and the equivalent dynamical equation in controllable canonical form

$$\dot{\vec{z}}(t) = \begin{pmatrix} 0 & 1 \\ -2 & -3 \end{pmatrix} \vec{z}(t) + \begin{pmatrix} 0 \\ 1 \end{pmatrix} u(t)$$

where $\vec{z}(t) = P^{-1}\vec{x}(t)$ and

$$P^{-1} = \begin{pmatrix} 1 & 2 \\ 1 & 3 \end{pmatrix}.$$

Let the desired closed-loop eigenvalues be $\lambda_1^D = -1 + i$ and $\lambda_2^D = -1 - i$. Then

$$(\lambda - \lambda_1^D)(\lambda - \lambda_2^D) = \lambda^2 + 2\lambda + 2$$

and $k_1 = 3 - 2 = 1$ and $k_2 = 2 - 2 = 0$. The gain matrices for the system in controllable canonical form and the original system are given by, respectively,

$$K = \begin{pmatrix} 0 & 1 \end{pmatrix},$$

$$KP^{-1} = \begin{pmatrix} 0 & 1 \end{pmatrix} \begin{pmatrix} 1 & 2 \\ 1 & 3 \end{pmatrix} = \begin{pmatrix} 1 & 3 \end{pmatrix}.$$

Remark 7.4.1 (Multi-Input Case, $u(\cdot) \in \mathbb{R}^m$, $m > 1$)
One approach is to change the multi-input problem into a single-input problem and then apply the result above. To this end, let us consider

$$\dot{\vec{x}}(t) = A\vec{x}(t) + B\vec{u}(t)$$

with state feedback law,

$$\vec{u}(t) = K\vec{x}(t),$$

where K is a constant $m \times n$ matrix. Let $\vec{d} \in \mathbb{R}^m$ and $\vec{k} \in \mathbb{R}^n$ and assume that

$$K = \vec{d}\vec{k}^T.$$

The closed-loop system is then given by

$$\dot{\vec{x}}(t) = A\vec{x}(t) + B\vec{d}\vec{k}^T\vec{x}(t)$$
$$= A\vec{x}(t) + \vec{b}\vec{k}^T\vec{x}(t),$$

where $\vec{b} = B\vec{d} \in \mathbb{R}^n$. We now note that if $(A, \vec{b})$ is controllable, then the method for the single-input case can be applied to find the "gain" $\vec{k}^T$. The problem then becomes: Given (A, B) controllable, can we find a vector $\vec{d} \in \mathbb{R}^m$ such that $(A, \vec{b})$, where $\vec{b} = B\vec{d}$, is controllable? The answer is positive if (A, B) is controllable and A is cyclic.

We note that since many $\vec{d}$ exist, the gain $\vec{d}\vec{k}^T$ is not unique. The matrix A is said to be cyclic if and only if the Jordan canonical form of A has one and only one Jordan block associated with each distinct eigenvalue. For example, the matrix

$$A = \left(\begin{array}{cc|c} 3 & 1 & 0 \\ 0 & 3 & 0 \\ \hline 0 & 0 & 3 \end{array}\right)$$

is not cyclic. Clearly, if A has distinct eigenvalues, then A is cyclic. For a discussion of this idea and of other methods to construct the gain matrix K for the multi-input case, including a method using controllable canonical form, see [6].

7.4.1 State Estimator (Luenberger Observer)

When we introduced state feedback, we assumed that all the state variables were available to be fed back to the system. In practice this assumption is not always met because

(i) all the state variables are not accessible to direct measurement, or

(ii) the number of measurement devices is limited (due to cost).

Thus, in order to utilize state feedback, we must reconstruct or obtain a good estimate of the state vector $\vec{x}$. We will reconstruct the state variable by using the available inputs and outputs of the system dynamics.

Consider the open-loop, time-invariant system

$$\dot{\vec{x}}(t) = A\vec{x}(t) + B\vec{u}(t)$$

$$\vec{y}(t) = C\vec{x}(t)$$

with the initial condition $\vec{x}(0) = \vec{x}_0$. We model the state estimator, $\hat{x}(t)$, with the same dynamic as the original system, so that

$$\dot{\hat{x}}(t) = A\hat{x}(t) + B\vec{u}(t) + G(\vec{y}(t) - \hat{y}(t))$$

$$\hat{y}(t) = C\hat{x}(t),$$

subject to $\hat{x}(0) = 0$ where $\vec{y}$ is the measured output and $\hat{y}(t)$ is the estimated output. The $n \times p$ matrix G is called the *observer gain matrix*. For stochastic systems, this is the *Kalman filter*.

We define the estimator error, $\vec{e}(t)$, as follows

$$\vec{e}(t) \equiv \hat{x}(t) - \vec{x}(t).$$

Then,

$$\begin{aligned}
\dot{\vec{e}}(t) &= \dot{\hat{x}}(t) - \dot{\vec{x}}(t) \\
&= A\hat{x}(t) + B\vec{u}(t) + G(\vec{y}(t) - \hat{y}(t)) - A\vec{x}(t) - B\vec{u}(t) \\
&= A\vec{e}(t) + GC(\vec{x}(t) - \hat{x}(t)) \\
&= (A - GC)\vec{e}(t)
\end{aligned}$$

and $\vec{e}(0) = \hat{x}(0) - \vec{x}(0) = 0 - x_0 = -x_0$. The problem, then, is to choose G such that $(A - GC)$ is asymptotically stable and $\lim_{t \to \infty} \vec{e}(t) = 0$ (i.e., $\lim_{t \to \infty}(\hat{x}(t) - \vec{x}(t)) = 0)$. Therefore, the question is whether we can we choose G so that the poles of $A - GC$ are anywhere we like in $\mathbb{C}$ (in particular, in the left-half complex plane). The answer is yes if (A, C) is observable.

Let $\gamma_1^D, \ldots, \gamma_n^D$ be the desired poles of $A - GC$ in the left-half plane of $\mathbb{C}$. Recall that the poles of $(A - GC)$ are the poles of $(A - GC)^T$, and observe that

$$(A - GC)^T = A^T + C^T(-G^T)$$
$$= \tilde{A} + \tilde{B}\tilde{K},$$

where $\tilde{A} = A^T$, $\tilde{B} = C^T$ and $\tilde{K} = -G^T$. We know that (A, C) is observable if and only if $(\tilde{A}, \tilde{B})$ is controllable. Thus, we can use any pole placement scheme to find the state variable gain $\tilde{K}$ that causes $\tilde{A} + \tilde{B}\tilde{K}$ to have the desired poles $\gamma_1^D, \ldots, \gamma_n^D$. Then, the state estimator gain matrix $G = -\tilde{K}^T$ will cause $A - GC$ to have the same desired poles.

Remark 7.4.2 The state estimator design is a pole placement design on $\tilde{A} = A^T$ and $\tilde{B} = C^T$.

7.4.2 Dynamic Output Feedback Compensator

Let $\hat{x}(t)$ be a state estimator. Consider the control law

$$\vec{u}(t) = K\hat{x}(t) + \vec{u}_r(t),$$

where u_r is the reference input and the state estimator

$$\dot{\hat{x}}(t) = A\hat{x}(t) + B\vec{u}(t) + G(\vec{y}(t) - \hat{y}(t))$$
$$\hat{y}(t) = C\hat{x}(t)$$

subject to the initial condition $\hat{x}(0) = 0$. This can be rewritten as

$$\vec{u}(t) = K\hat{x}(t) + \vec{u}_r(t)$$

$$\dot{\hat{x}}(t) = L\hat{x}(t) + G\vec{y}(t) + B\vec{u}_r(t),$$

where $L = A + BK - GC$.

We want to find the closed-loop poles of the system. We have

$$\dot{\vec{x}}(t) = A\vec{x}(t) + B\vec{u}(t),$$

where

$$\vec{u}(t) = K\hat{x}(t) + \vec{u}_r(t)$$
$$= K(\vec{x}(t) + \vec{e}(t)) + \vec{u}_r(t)$$

and $\dot{\vec{e}}(t) = (A - GC)\vec{e}(t)$. Therefore,

$$\dot{\vec{x}}(t) = A\vec{x}(t) + BK\vec{x}(t) + BK\vec{e}(t) + Bu_r(t)$$
$$= (A + BK)\vec{x}(t) + BK\vec{e}(t) + Bu_r(t).$$

Consequently, the closed-loop system is given by the composite system:

$$\begin{pmatrix} \dot{\vec{x}}(t) \\ \dot{\vec{e}}(t) \end{pmatrix} = \begin{pmatrix} A + BK & BK \\ 0 & A - GC \end{pmatrix} \begin{pmatrix} \vec{x}(t) \\ \vec{e}(t) \end{pmatrix} + \begin{pmatrix} B \\ 0 \end{pmatrix} u_r(t).$$

It then follows that the poles of the closed-loop system are those of $A + BK$ (full state feedback design) and $A - GC$ (state estimator design). This is known as the *deterministic separation principle*. The dynamic output compensator is summarized in Figure 7.8.

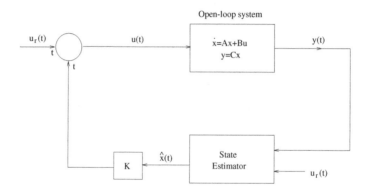

FIGURE 7.8:   Dynamic output compensator.

Given the desired poles for the system $\lambda_1^D, \ldots, \lambda_n^D$, the state estimator is designed using the following algorithm.

Algorithm

1. If (A, B) is controllable, we can find a matrix K such that $A + BK$ has the desired poles $\lambda_1^D, \cdots, \lambda_n^D$.

2. Use the control law
$$\vec{u}(t) = K\hat{x}(t) + \vec{u}_r(t).$$

   That is, use the estimated states as though they were the actual states.

3. The state estimator is given by
$$\dot{\hat{x}}(t) = A\hat{x}(t) + B\vec{u}(t) + G(\vec{y}(t) - \hat{y}(t))$$

$$\hat{y}(t) = C\hat{x}(t)$$

subject to the initial condition $\hat{x}(0) = 0$. The estimator error is

$$\dot{\vec{e}}(t) = (A - GC)\vec{e}(t)$$

with $\vec{e}(0) = -x_0$. If (A, C) is observable, we can find a matrix G such that $A - GC$ has the desired poles $\gamma_1^D, \cdots, \gamma_n^D$; these are chosen by the designer. The rule of thumb is for the estimator error poles to be placed slightly to the left of the controlled system poles, i.e. $Re(\gamma_i^D) < Re(\lambda_i^D)$, for $i = 1, 2, \cdots, n$.

The closed-loop system will have the poles of $A + BK$ (i.e., the desired poles $\lambda_1^D, \cdots, \lambda_n^D$) plus the estimator error poles of $A - GC$ (i.e., $\gamma_1^D, \cdots, \gamma_n^D$). The designer can achieve any set of poles $\{\lambda_1^D, \cdots, \lambda_n^D\}$ and $\{\gamma_1^D, \cdots, \gamma_n^D\}$ in $\mathbb{C}$ by *separately* choosing the gains K and G (Separation Principle).

Example 7.4.5 Let

$$\dot{\vec{x}}(t) = A\vec{x}(t) + B\vec{u}(t)$$
$$\vec{y}(t) = C\vec{x}(t),$$

where

$$A = \begin{pmatrix} 0 & 1 & 0 \\ 0 & 0 & 1 \\ 0 & -1 & 0 \end{pmatrix}, \quad B = \begin{pmatrix} 0 \\ 0 \\ 1 \end{pmatrix}, \quad \text{and} \quad C = \begin{pmatrix} 1 & -1 & 0 \end{pmatrix}.$$

We want to design a dynamic output feedback compensator with desired poles $\lambda_1^D = -1, \lambda_2^D = -1+j, \lambda_3^D = -1-j$, and state estimator poles $\gamma_1^D = -2, \gamma_2^D = -2+j$, and $\gamma_3^D = -2-j$. Note that the system is controllable, since

$$\rho([B|AB|A^2B]) = \rho \begin{pmatrix} 0 & 0 & 1 \\ 0 & 1 & 0 \\ 1 & 0 & -1 \end{pmatrix} = 3,$$

and the system is observable since

$$\rho \begin{pmatrix} C \\ CA \\ CA^2 \end{pmatrix} = \rho \begin{pmatrix} 1 & -1 & 0 \\ 0 & 1 & -1 \\ 0 & 1 & 1 \end{pmatrix} = 3.$$

The open-loop poles are $\lambda_1 = 0, \lambda_2 = j$, and $\lambda_3 = -j$. Note in Figure 7.9 that the system is not asymptotically stable.

Using Separation Principle, we design the dynamic output feedback compensator as follows:

(a) First, we want to design a gain matrix K so that $A + BK$ has the desired poles $\lambda_1^D, \lambda_2^D, \lambda_3^D$. Such a matrix is

$$K = \begin{pmatrix} -2 & -3 & -3 \end{pmatrix}.$$

Note that this is a single-input system in controllable canonical form.

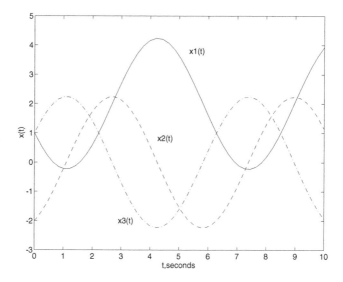

FIGURE 7.9:   The uncontrolled system ($u \equiv 0$.)

(b) Next, we design a matrix G such that $A - GC$ has poles $\gamma_1^D, \gamma_2^D, \gamma_3^D$. It is given by

$$G = \begin{pmatrix} 14 \\ 8 \\ -4 \end{pmatrix}.$$

Note that (A^T, C^T) is not in controllable canonical form. In this case, we can either transform the system to canonical form or use a direct design method.

The closed-loop system and the estimator errors are plotted in Figures 7.10 and 7.11, respectively. Note that the closed-loop system is asymptotically stable. The state estimator, $\hat{x}(t)$, is plotted in Figure 7.12.

Now, suppose we want to design a dynamic feedback compensator for this system with the desired poles shifted further to the left in the complex plane, say $\lambda_1^D = -4, \lambda_2^D = -4 + j, \lambda_3^D = -4 - j$, and state estimator poles $\gamma_1^D = -5, \gamma_2^D = -5 + j$, and $\gamma_3^D = -5 - j$. We design a gain matrix K so that $A + BK$ has the desired poles $\lambda_1^D, \lambda_2^D, \lambda_3^D$:

$$K = \begin{pmatrix} -68 & -48 & -12 \end{pmatrix}.$$

Similarly, we design a matrix G such that $A - GC$ has poles $\gamma_1^D, \gamma_2^D, \gamma_3^D$:

$$G = \begin{pmatrix} 110 \\ 95 \\ 20 \end{pmatrix}.$$

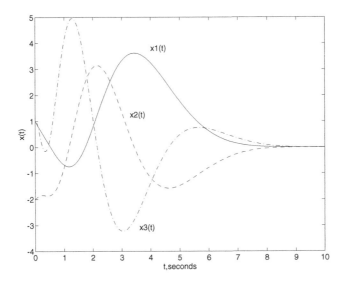

FIGURE 7.10: The state vector, $x(t)$, of the closed-loop system with $K = (-2 \ -3 \ -3)$ and $G = (\ 14 \ 8 \ -4)^T$.

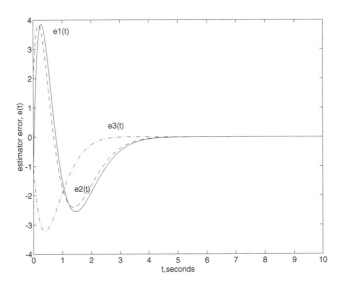

FIGURE 7.11: The estimator error, $e(t)$, of the closed-loop system with $K = (-2 \ -3 \ -3)$ and $G = (\ 14 \ 8 \ -4)^T$.

Figures 7.13 and 7.14 illustrate that both the state and the estimator error go to zero faster than in the previous case, and hence, so does the state estimator

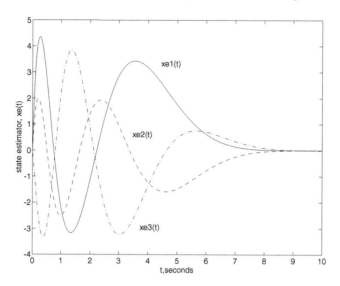

FIGURE 7.12: The state estimator, $\hat{x}(t)$, with $K = (-2 \ -3 \ -3)$ and $G = (\ 14\ 8\ -4)^T$. The label $xe1(t)$ denotes $\hat{x}_1(t)$, etc.

(Figure 7.15).

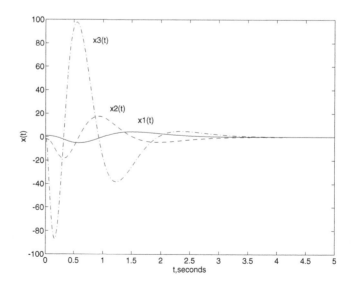

FIGURE 7.13: The state vector, $x(t)$, with $K = (-68 \ -48 \ -12)$ and $G = (110\ 95\ 20)^T$.

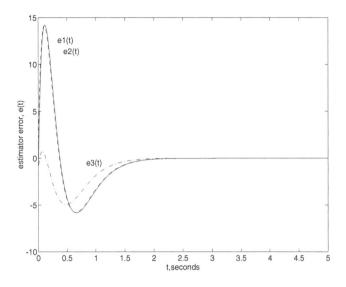

FIGURE 7.14: The estimator error, $e(t)$, with $K = (-68 \ -48 \ -12)$ and $G = (110 \ \ 95 \ \ 20)^T$.

Note, too, that the matrices K and G are greater in norm than those of the previous case; that is, as we shift the desired poles farther to the left in the complex plane, we gain in the rate of decay but we incur additional penalty in the process (more control). In the next section, we will discuss methods to find the optimal control that minimizes a cost criteria.

7.5 Linear Quadratic Regulator Theory

Consider the linear time-invariant control system

$$\dot{\vec{x}}(t) = A\vec{x}(t) + B\vec{u}(t) \tag{7.33}$$

with $\vec{u}(t) = K\vec{x}(t)$. Then the closed-loop eigenvalues of

$$\dot{\vec{x}}(t) = (A + BK)\vec{x}(t) \tag{7.34}$$

can be arbitrarily located in the complex plane if (A, B) is controllable. However, Example 7.4.5 in the last section demonstrated that the faster we make the system converge to the zero state, the larger the amplitude of K is (and, therefore, more input or control will be required). This leads us to the following formulation of an optimal control problem: Find a control

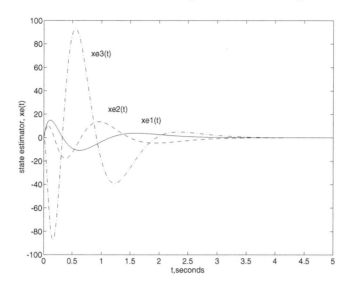

FIGURE 7.15: The state estimator, $\hat{x}(t)$, with $K = (-68 \;\; -48 \;\; -12)$ and $G = (110 \;\; 95 \;\; 20)^T$. The label $xe1(t)$ denotes $\hat{x}_1(t)$, etc.

$u \in L_2(t_0, \infty; R^m)$ to minimize

$$J(\vec{x}_0, \vec{u}) = \frac{1}{2} \int_{t_0}^{\infty} (\vec{x}^T Q \vec{x} + \vec{u}^T R \vec{u}) \, dt, \tag{7.35}$$

where $\vec{x} \in R^n$, $\vec{u} \in R^m$, $Q \in R^{n \times n}$ is symmetric positive semidefinite (SPSD), and $R \in R^{m \times m}$ is symmetric positive definite (SPD). Associated with the performance index (7.35) are the linear dynamics described by (7.33). This optimal control problem is known as the linear quadratic regulator problem [1]. It is not always possible to solve this problem as stated. For example, consider

$$\dot{\vec{x}}(t) = \begin{pmatrix} 1 & 0 \\ 0 & 1 \end{pmatrix} \vec{x}(t) + \begin{pmatrix} 0 \\ 1 \end{pmatrix} u(t)$$

with initial condition $\vec{x}(0) = (1, 0)^T$ and

$$J(\vec{x}_0, u) = \int_0^{\infty} (x_1^2 + x_2^2 + u^2) \, dt.$$

We note that $x_1(t) = e^t$ for any arbitrarily control function $u(t)$. Also, since $\dot{x}_2 = x_2 + u$ with $x_2(0) = 0$, it is easy to see that the optimal control $u^\star = 0$ which, then, implies that the corresponding optimal trajectory $x_2^\star = 0$. Hence,

$$J(\vec{x}_0, u^\star) = \int_0^{\infty} e^{2t} \, dt = \infty.$$

This example shows that the optimal cost is not finite because of the following three reasons:

(i) the state x_1 is uncontrollable,

(ii) the uncontrollable state is unstable,

(iii) the unstable state is part of the cost functional.

This difficulty would not arise if we assumed that (A, B) is controllable. In fact, if (A, B) is controllable and $Q > 0$, it has been shown [1] that

(a) The unique optimal control is given by $\vec{u}^*(t) = -K\vec{x}^*(t)$, where $K = R^{-1}B^T\Pi$, Π is the unique positive definite matrix solution to the algebraic Riccati equation

$$A^T\Pi + \Pi A - \Pi BR^{-1}B^T\Pi + Q = 0$$

and $\vec{x}^*(t)$ is the solution to the closed-loop system

$$\dot{\vec{x}}^*(t) = (A - BR^{-1}B^T\Pi)\vec{x}^*(t), \qquad \vec{x}^*(t_0) = \vec{x}_0.$$

(b) Moreover, the matrix $(A - BR^{-1}B^T\Pi)$ has all eigenvalues with negative real parts (hence, the closed-loop system is asymptotically stable) and the optimal cost is given by

$$J^*(\vec{x}_0, \vec{u}^*) = \frac{1}{2}\vec{x}_0^T\Pi\vec{x}_0.$$

The above optimal control is in full state feedback form. That is, we assume that all state variables are available to be fed back. Since, in general, we do not have full state information, we will make use of a state estimator to formulate the controller. From [1], the optimal state estimator (observer) takes the form

$$\dot{\hat{x}}(t) = A\hat{x}(t) + B\vec{u}(t) + G(\vec{y}(t) - \hat{y}(t)) \tag{7.36}$$
$$\hat{y}(t) = C\hat{x}(t), \tag{7.37}$$

where $\vec{y}$ is the measured output. The optimal observer gain G is given by $G = \Sigma C^T\hat{R}^{-1}$ and Σ solves the algebraic Riccati equation for the dual system

$$\Sigma A^T + A\Sigma - \Sigma C^T\hat{R}^{-1}C\Sigma + \hat{Q} = 0.$$

Playing a similar role to that of the matrices Q and R in the optimal state feedback problem, the symmetric positive semidefinite matrix $\hat{Q}$ and the symmetric positive definite matrix $\hat{R}$ are design criteria for the optimal state estimator.

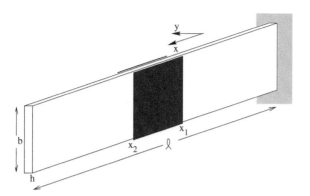

FIGURE 7.16:   Cantilever beam with piezoceramic patches.

7.6   Beam Vibrational Control: Real-Time Feedback Control Implementation

In this section we present the application of the control theory discussed in previous sections to the control of transverse beam vibrations. The beam structure is an aluminum cantilever beam in a "smart structure" paradigm (see Chapter 6). Using the same notation as in [15] we denote the beam length, width, thickness, density, Young's modulus and Kelvin-Voigt damping by ℓ, b, h, ρ, E and c_D, respectively. The origin is taken to be at the clamped edge of the beam, and the axial direction is denoted by the x-axis (see Figure 7.16). A pair of identical piezoceramic patches are bonded on opposite sides of the beam with edges located at x_1 and x_2. Passive patch contributions arising from material changes due to the presence of the patches are included in the model. Patch parameters are denoted by the subscript pe, thus, the patch thickness, width, density, Young's modulus and Kelvin-Voigt damping are given by $h_{pe}, b_{pe}, \rho_{pe}, E_{pe}$ and c_{Dpe}, respectively. We denote the piezoelectric constant relating mechanical strain and applied electric field by d_{31}. Finally, we denote the transverse displacement by y and the voltages applied to the front and back patches by V_1 and V_2, respectively.

As derived in [3], the transverse (or bending) equation of the beam is given in terms of resultant moments bM_x by

$$\tilde{\rho}\frac{\partial^2 y}{\partial t^2} - \frac{\partial^2(bM_x)}{\partial x^2} + \frac{\partial^2(bM_x)_{pe}}{\partial x^2} = f. \qquad (7.38)$$

Passive patch contributions are incorporated in the model above and hence the linear mass density $\tilde{\rho}(x) = \rho hb + 2b\rho_{pe}h_{pe}\chi_{pe}(x)$ is piecewise constant. The characteristic function $\chi_{pe}(x)$ employed to isolate patch contributions is defined by

$$\chi_{pe}(x) = \begin{cases} 1 \,, & x_1 \leq x \leq x_2 \\ 0 \,, & \text{otherwise.} \end{cases} \tag{7.39}$$

Incorporating both internal damping and material changes due to the presence of the patches, the internal moment resultant bM_x has the form

$$bM_x(t, x) = -\widetilde{EI}(x)\frac{\partial^2}{\partial x^2}y(t, x) - \widetilde{c_D I}(x)\frac{\partial^3}{\partial x^2 \partial t}y(t, x), \tag{7.40}$$

where

$$\widetilde{EI}(x) = E\frac{h^3 b}{12} + \frac{2b}{3}E_{pe}a_3\chi_{pe}(x), \quad \widetilde{c_D I}(x) = c_D\frac{h^3 b}{12} + \frac{2b}{3}c_{Dpe}a_3\chi_{pe}(x), \tag{7.41}$$

and $a_3 = (h/2 + h_{pe})^3 - h^3/8$ (we refer the reader to Chapter 3 of [3] for details regarding patch contributions to the internal moment resultant). When voltages are applied to the front and back patches, the induced external moment $(bM_x)_{pe}$ is given by

$$(bM_x)_{pe}(t, x) = \frac{1}{2}E_{pe}bd_{31}(h + h_{pe})\chi_{pe}(x)[V_1(t) - V_2(t)]. \tag{7.42}$$

External transverse forces acting on the beam are modeled by the function $f(t, x)$. Cantilever boundary conditions are given by

$$y(t, 0) = \frac{\partial y}{\partial x}(t, 0) = 0, \qquad M_x(t, \ell) = \frac{\partial}{\partial x}M_x(t, \ell) = 0, \tag{7.43}$$

and initial conditions are denoted by

$$y(0, x) = y_0(x), \qquad \frac{\partial y}{\partial t}(0, x) = y_1(x). \tag{7.44}$$

From previous discussions in Chapter 6, to approximate the solution to the strong form (7.38) we first write the weak formulation of (7.38) as

$$\int_0^\ell \rho h b \frac{\partial^2 y(t, x)}{\partial t^2}\phi(x)dx + \int_{x_1}^{x_2} 2b\rho_{pe}h_{pe}\frac{\partial^2 y(t, x)}{\partial t^2}\phi(x)dx$$

$$+ \int_0^\ell EI\frac{\partial^2 y(t, x)}{\partial x^2}\frac{\partial^2 \phi(x)}{\partial x^2}dx + \int_{x_1}^{x_2}\frac{2b}{3}E_{pe}a_3\frac{\partial^2 y(t, x)}{\partial x^2}\frac{\partial^2 \phi(x)}{\partial x^2}dx$$

$$+ \int_0^\ell c_D I\frac{\partial^3 y(t, x)}{\partial x^2 \partial t}\frac{\partial^2 \phi(x)}{\partial x^2}dx + \int_{x_1}^{x_2}\frac{2b}{3}c_{Dpe}a_3\frac{\partial^3 y(t, x)}{\partial x^2 \partial t}\frac{\partial^2 \phi(x)}{\partial x^2}dx \tag{7.45}$$

$$= \int_{x_1}^{x_2}\frac{1}{2}E_{pe}bd_{31}(h + h_{pe})\Big(V_1(t) - V_2(t)\Big)\frac{\partial^2 \phi(x)}{\partial x^2}dx$$

$$+ \int_0^\ell f(t, x)\phi(x)dx,$$

and approximate the solution to the weak form (7.45) by performing the Galerkin expansion

$$y^{\mathcal{N}}(t,x) = \sum_{i=1}^{\mathcal{N}} z_i(t)\mathcal{B}_i(x), \qquad (7.46)$$

where $\mathcal{B}_i$ are the cubic splines functions (6.63). This yields the matrix system approximating (7.45) of the form

$$(M + M_{pe})\ddot{z}(t) + (D + D_{pe})\dot{z}(t) + (KE + KE_{pe})z(t) = \tilde{F}(t) + \tilde{B}u(t)$$

$$z(0) = z_0, \qquad \dot{z}(0) = z_1,$$

$$(7.47)$$

where $z(t) = [z_1(t), \ldots, z_{\mathcal{N}}(t)]^T$ is the vector of coefficients. The matrices in (7.47) are defined by

$$[M]_{k,l} = \rho h b \int_0^\ell \mathcal{B}_l(x)\mathcal{B}_k(x)dx, \quad [M_{pe}]_{k,l} = 2b\rho_{pe}h_{pe}\int_{x_1}^{x_2}\mathcal{B}_l(x)\mathcal{B}_k(x)dx$$

$$[KE]_{k,l} = EI\int_0^\ell \mathcal{B}_l''(x)\mathcal{B}_k''(x)dx, \quad [KE_{pe}]_{k,l} = \frac{2b}{3}E_{pe}a_3\int_{x_1}^{x_2}\mathcal{B}_l''(x)\mathcal{B}_k''(x)dx$$

$$[D]_{k,l} = c_D I\int_0^\ell \mathcal{B}_l''(x)\mathcal{B}_k''(x)dx, \quad [D_{pe}]_{k,l} = \frac{2b}{3}c_{Dpe}a_3\int_{x_1}^{x_2}\mathcal{B}_l''(x)\mathcal{B}_k''(x)dx$$

$$[\tilde{F}]_k(t) = \int_0^\ell f(x,t)\mathcal{B}_k(x)dx, \quad [\tilde{B}]_{k,1} = \tfrac{1}{2}E_{pe}bd_{31}(h+h_{pe})\int_{x_1}^{x_2}\mathcal{B}_k''(x)dx$$

$$u(t) = [V_1(t)\,,\,V_2(t)]^T, \quad [\tilde{B}]_{k,2} = -\tfrac{1}{2}E_{pe}bd_{31}(h+h_{pe})\int_{x_1}^{x_2}\mathcal{B}_k''(x)dx.$$

$$(7.48)$$

The first order reformulation of (7.47) is given by

$$\dot{w}(t) = Aw(t) + Bu(t) + F(t)$$

$$w(0) = w_0 = [z_0, z_1]^T,$$

$$(7.49)$$

where $w = (z, \dot{z})^T$,

$$A = \begin{bmatrix} I & 0 \\ 0 & M + M_{pe} \end{bmatrix}^{-1} \begin{bmatrix} 0 & I \\ -(KE + KE_{pe}) & -(D + D_{pe}) \end{bmatrix}, \qquad (7.50)$$

$$B = \begin{bmatrix} I & 0 \\ 0 & M + M_{pe} \end{bmatrix}^{-1} \begin{bmatrix} 0 \\ \tilde{B} \end{bmatrix} \quad \text{and} \quad F(t) = \begin{bmatrix} I & 0 \\ 0 & M + M_{pe} \end{bmatrix}^{-1} \begin{bmatrix} 0 \\ \tilde{F}(t) \end{bmatrix}. \quad (7.51)$$

Detailed discussion of the Galerkin approximation for the beam model (7.38) and some theoretical considerations including convergence analysis can be found in [3, 15]. Finally, the structure of the observation matrix C in the output equation $y(t) = Cw(t)$ depends on the sensor employed in the experiment. In our model, the sensor is a proximity probe located at the back of the beam sensing displacements at the point x_{ob}. The observation matrix C is thus of the form

$$C = \left[B_1(x_{ob}), \cdots, B_N(x_{ob}), \underbrace{0, \cdots, 0}_{N} \right],$$

where the $B_i(x_{ob})$'s are the basis functions evaluated at x_{ob}.

In real-time implementation, the signals from the sensors are digitized and the real-time processor can only perform at a discrete sample rate Δt. Thus, the state estimator equation

$$\dot{\hat{w}}(t) = A\hat{w}(t) + Bu(t) + F(y(t) - \hat{y}(t)) \tag{7.52}$$

$$\hat{y}(t) = C\hat{w}(t) \tag{7.53}$$

can only be evolved in time in discrete time steps, and the control voltage can be only computed at this rate. The numerical ODE approximation method to solve the state estimator equation (7.52) must satisfy the following criteria:

(i) the control $u(t_j)$ must be calculated before the arrival of the data at the next time step $t_{j+1} = t_j + \Delta t$,

(ii) the method must be sufficiently accurate to resolve system dynamics and

(iii) since the ODE systems are often stiff, the method must be A-stable or α-stable (see, e.g., [13]).

We chose a modified backward Euler method given in Chapter 8.2.1 of [3]. The fast sample rate (and hence small Δt) at which we can carry out the experiment allows the use of this method. An A-stable modified backward Euler method integrating the state estimator at time t_{j+1} is given by

$$\begin{aligned} \hat{w}_{j+1} &= (I - \Delta t A_c)^{-1}\hat{w}_j + \Delta t(I - \Delta t A_c)^{-1}Fy(t_j) \\ &= R(A_c)\hat{w}_j + \Delta t R(A_c)Fy(t_j), \end{aligned} \tag{7.54}$$

where $A_c = A - BK - FC$, $R(A_c) = (I - \Delta t A_c)^{-1}$ and the constant time step is $\Delta t = t_{j+1} - t_j$. Note that the method is modified from standard backward Euler methods since the observation $y(t_j)$ at future time steps are not available. We now summarize the discrete-time algorithm in the following Real-Time Control Algorithm, which is essentially Algorithm 8.5 in [3].

Real-Time Control Algorithm

(a) *Offline*

(i) Construct matrices $A, B, C, Q, R, \hat{Q}, \hat{R}$.

(ii) Solve the Riccati equations for Π and Σ.

TABLE 7.1:   Beam and patch parameters.

Beam	Patch
$\ell = 0.286$ m	$h_{pe} = 5.3 \times 10^{-4} 4$ m
$h = 0.001$ m	$\rho_{pe} = 7.45 \times 10^3$ kg/m^3
$b = 0.2543$ m	$E_{pe} = 6.4 \times 10^{10}$ N/m^2
$\rho = 3.438 \times 10^3$ kg/m^3	$c_{Dpe} = 3.96 \times 10^5$ Ns/m^2
$E = 7.062 \times 10^{10}$ N/m^2	$d_{31} = 262 \times 10^{-12}$ m/V
$c_D = 1.04 \times 10^6$ Ns/m^2	$x_1 = 0.02041$ m
$\hat{x} = 0.11076$ m	$x_2 = 0.04592$ m

(iii) Construct $K = R^{-1}B^T\Pi$, $F = \Sigma C^T \hat{R}^{-1}$, and $A_c = A - BK - FC$.

(iv) Construct $R(A_c) = (I - \Delta t A_c)^{-1}$ and $R(A_c)F = (I - \Delta t A_c)^{-1}F$.

(b) *Online*

(i) Collect observation $y(t_j)$.

(ii) Time stepping the discrete compensator system

$$\hat{w}_{j+1} = R(A_c)\hat{w}_j + \Delta t R(A_c)Fy(t_j).$$

(iii) Calculate the voltage $u(t_{j+1}) = -K\hat{w}_{j+1}$.

In Table 7.1, we report the dimensions and parameters of our experimental beam structure depicted in Figure 7.17. The aluminum beam parameters ρ, E, c_D and the lead zirconate titanate piezoceramic patch parameters $\rho_{pe}, E_{pe}, c_{Dpe}, d_{31}$ were obtained from the manufacturers.

Numerical simulations were performed to obtain reasonable values of the control parameter matrices $Q, R, \hat{Q}$ and $\hat{R}$. We sought parameters leading to maximum control voltages within the $\pm 100V$ range of the patches while at the same time providing good attenuation. The matrices employed were of the form

$$Q = d_1 \begin{bmatrix} KE + KE_{pe} & 0 \\ 0 & M + M_{pe} \end{bmatrix}, \quad R = r_1 I^{p \times p}, p = 1$$

$$\hat{Q} = \hat{d}_1 \begin{bmatrix} I^N & 0 \\ 0 & I^N \end{bmatrix}, \quad\quad\quad\quad \hat{R} = \hat{r}_1 I^{s \times s}, s = 1, \tag{7.55}$$

where p is the number of actuators, s is the number of sensors, $d_1 = 2 \times 10^8$, $r_1 = 0.98$, $\hat{d}_1 = 1 \times 10^3$, and $\hat{r}_1 = 1$.

In Figure 7.18, we present a diagram of the experimental setup and implementation of the online component of the Real-Time Control Algorithm. Voltage spikes to the back patch (to excite the beam) were generated by a DS1103 dSpace control system. The excitation signal was low pass filtered (i.e., only the low frequency signal is retained) and amplified before being

FIGURE 7.17:   Experimental beam with piezoceramic patches.

applied to the back patch. The voltage spike was amplified so as to produce 90 volts at the peak. A proximity probe located at the back of the beam at $x = \hat{x} = 0.11076m$ was used to measure displacements and the observation readings were digitized through one analog to digital channel of the dSpace hardware. This observation signal enters the online component of the Real-Time Control Algorithm as $y(t_j)$. By employing the discrete modified backward Euler method (7.54), the state estimator $\hat{w}(t_{j+1})$ was obtained and multipled with the gain matrix K to produce the control voltage. The control signal was then low pass filtered and amplified before being sent to the front patch. A constant discrete time rate of $\Delta t = 10^{-4}s$ was employed in running the real-time processor.

In Figure 7.19 and 7.20, we report the uncontrolled and controlled displacements and the control voltage, respectively. Note that the control system has basically attenuated the displacements after one second.

Project: Control Design

Consider the inverted pendulum system mounted on a motor-driven cart as shown in Figure 7.21. Here, we consider only the two-dimensional problem where the pendulum moves only in the plane of the paper.

The inverted pendulum is unstable in that it may fall over any time in any direction. It is desired to keep the pendulum upright in the presence of

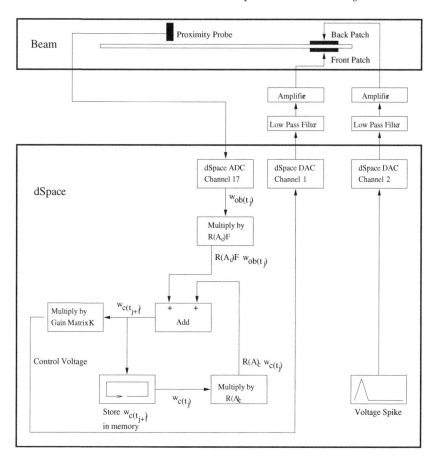

FIGURE 7.18:   Experimental setup and implementation of online component of the Real-Time Control Algorithm.

disturbances (such as a gust of wind acting on the mass or an unexpected
force applied to the cart). The slanted pendulum can be brought back to the
vertical position when appropriate control force $u(t)$ is applied to the cart. At
the end of each control process, it is also desired to bring the cart back to the
origin position $x = 0$, the reference position.

Design a control system such that, given any initial conditions (caused by
disturbances), the pendulum can be brought back to the vertical position and
also the cart can be brought back to the reference position ($x = 0$). Thus,
this is a regulator problem and the controller can be designed using the linear
quadratic regulator (LQR) technique. We assume the following numerical
values for M, m and l:

$$M = 2 \text{ kg}, m = 0.1 \text{ kg}, l = 0.5 \text{ m},$$

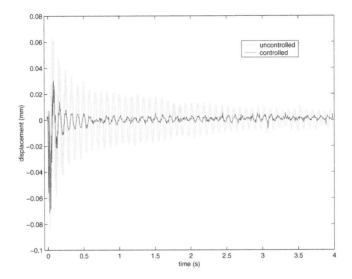

FIGURE 7.19: Uncontrolled and controlled displacements at $x_{ob} = 0.11075m$.

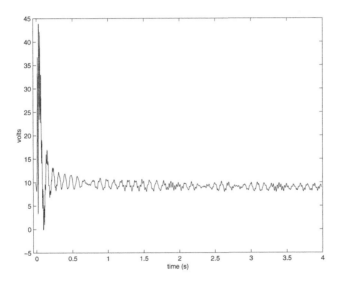

FIGURE 7.20: Control voltages.

and that the pendulum mass is concentrated at the top of the rod and the rod itself is essentially massless.

(i) *A mathematical model.*

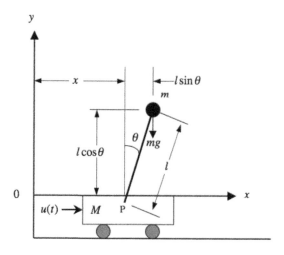

FIGURE 7.21: The inverted pendulum.

To derive the equations of motion for the system, consider the free body diagram of the inverted pendulum system as depicted in Figure 7.22.

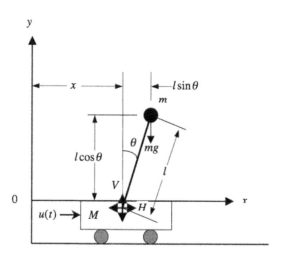

FIGURE 7.22: Free body diagram of the inverted pendulum.

The rotational motion of the pendulum rod about the center of gravity of the mass m is described by

$$I\ddot{\theta} = Vl\sin(\theta) - Hl\cos(\theta), \tag{7.56}$$

where I is the moment of inertia of the pendulum rod about its center of gravity. The horizontal motion of the mass m is given by

$$m\frac{d^2}{dt^2}(x + l\sin(\theta)) = H \tag{7.57}$$

and the vertical motion of the mass m is given by

$$m\frac{d^2}{dt^2}(l\cos(\theta)) = V - mg. \tag{7.58}$$

Finally, the horizontal motion of the cart is described by

$$M\frac{d^2x}{dt^2} = u - H. \tag{7.59}$$

These equations are nonlinear because of the nonlinearities in (7.56), (7.57) and (7.58) and the coupling between θ and x. In order to apply the linear control theory presented in this chapter we will now proceed to linearize the above equations. We assume that the angle θ is small so that $\sin(\theta) \approx \theta$ and $\cos(\theta) \approx 1$. This assumption reduces the equations (7.56-7.59) to

$$I\ddot{\theta} = Vl\theta - Hl, \tag{7.60}$$
$$m(\ddot{x} + l\ddot{\theta}) = H, \tag{7.61}$$
$$0 = V - mg, \tag{7.62}$$
$$M\ddot{x} = u - H. \tag{7.63}$$

From equations (7.61) and (7.63), we obtain

$$(M + m)\ddot{x} + ml\ddot{\theta} = u. \tag{7.64}$$

From equations (7.60) and (7.62), we have

$$I\ddot{\theta} = mgl\theta - Hl = mgl\theta - l(m\ddot{x} + ml\ddot{\theta})$$

or

$$(I + ml^2)\ddot{\theta} + ml\ddot{x} = mgl\theta. \tag{7.65}$$

For this project, we now assume that the moment of inertia $I = 0$. Therefore, equations (7.64) and (7.65) can be modified to

$$Ml\ddot{\theta} = (M + m)g\theta - u, \tag{7.66}$$
$$M\ddot{x} = u - mg\theta. \tag{7.67}$$

We note that equations (7.66) and (7.67) describe the motion of the inverted-pendulum-on-the-cart system (under the assumption that the angle θ is small). They constitute a mathematical model of the system. In order to design the feedback control law, we now rewrite the above equations (7.66) and (7.67) as a system of first order differential equations. To begin we define the state variables x_1, x_2, x_3 and x_4 by

$$x_1 = \theta, x_2 = \dot{\theta}, x_3 = x, x_4 = \dot{x}.$$

Note that the angle θ indicates the rotation of the pendulum rod about the point P and x is the location of the cart. We consider θ and x as the outputs of the system. That is,

$$\vec{y}(t) = \begin{bmatrix} x_1 \\ x_3 \end{bmatrix}.$$

In terms of vector-matrix equations, the state and output equations are described by

$$\begin{bmatrix} \dot{x}_1 \\ \dot{x}_2 \\ \dot{x}_3 \\ \dot{x}_4 \end{bmatrix} = \begin{bmatrix} 0 & 1 & 0 & 0 \\ \frac{M+m}{Ml}g & 0 & 0 & 0 \\ 0 & 0 & 0 & 1 \\ -\frac{m}{M}g & 0 & 0 & 0 \end{bmatrix} \begin{bmatrix} x_1 \\ x_2 \\ x_3 \\ x_4 \end{bmatrix} + \begin{bmatrix} 0 \\ -\frac{1}{Ml} \\ 0 \\ \frac{1}{M} \end{bmatrix} u,$$

$$\vec{y}(t) = \begin{bmatrix} y_1 \\ y_2 \end{bmatrix} = \begin{bmatrix} 1 & 0 & 0 & 0 \\ 0 & 0 & 1 & 0 \end{bmatrix} \begin{bmatrix} x_1 \\ x_2 \\ x_3 \\ x_4 \end{bmatrix}.$$

Using numerical values for M, m and l as given previously we obtain

$$\dot{\vec{x}} = A\vec{x} + Bu, \tag{7.68}$$
$$\vec{y} = C\vec{x}, \tag{7.69}$$

where

$$A = \begin{bmatrix} 0 & 1 & 0 & 0 \\ 20.601 & 0 & 0 & 0 \\ 0 & 0 & 0 & 1 \\ -0.4905 & 0 & 0 & 0 \end{bmatrix}, B = \begin{bmatrix} 0 \\ -1 \\ 0 \\ 0.5 \end{bmatrix},$$

$$C = \begin{bmatrix} 1 & 0 & 0 & 0 \\ 0 & 0 & 1 & 0 \end{bmatrix}.$$

(ii) *Model Analysis and State Feedback Control Design.*

1. When $u = 0$, show that the zero equilibrium state is unstable.

2. Show that the system is controllable and observable.

3. For $u = 0$ and $x(0) = [0.1, 0, 0, 0]^T$, compute and plot the solution trajectories. Comment on your solution curves.

4. Use the LQR formulation (use MATLAB routine `lqr`) to determine a full state observer (estimator) and a stabilizing linear state feedback control law. For the full state observer, use $\hat{Q} = 50I$ and $\hat{R} = 2I$, where I is the identity matrix. For the state feedback control law, use $Q = 10I$ and $R = 2I$. Plot the closed-loop system states, the feedback control, and the state estimator when $x(0)$ is the same as in part 3. above.

5. Now, let $\hat{Q} = 500I$ and $Q = 100I$, repeat part 4. and comment on the effects of the choices for the weights on the state trajectories of the closed loop system and on the control.

References

[1] B.D.O. Anderson and J.B. Moore, *Linear Optimal Control*, Prentice Hall, Englewood Cliffs, 1971.

[2] P.J. Antsaklis and A.N. Michel, *Linear Systems*, The McGraw-Hill Companies, Inc., New York, 1997.

[3] H.T. Banks, R.C. Smith and Y. Wang, *Smart Material Structures: Modeling, Estimation and Control*, Masson/John Wiley, 1996.

[4] S. Barnett and R.G. Cameron, *Introduction to Mathematical Control Theory*, Oxford Applied Mathematics and Computing Science Series, 2nd ed., 1985.

[5] W.L. Brogran, *Modern Control Theory*, Prentice Hall, Englewood Cliffs, 3rd ed., 1991.

[6] Chi-Tsong Chen, *Linear System Theory and Design*, Oxford University Press, Inc., New York, 1999.

[7] R. Datko, Not all feedback stabilized hyperbolic systems are robust with respect to small time delays in their feedbacks, *SIAM J. Control and Optimization*, **26**(3), 1988, pp. 697–713.

[8] R. Datko, J. Laguese and M.P. Polis, An Example on the effect of time delays in boundary feedback stabilization of wave equations, *SIAM J. Control and Optimization*, **24**(1), 1986, pp. 152–156.

[9] E.J. Davison, On pole assignment in multivariable linear systems, *IEEE Trans. on AC*, **AC-13**(6), 1968, pp. 747–748.

[10] Jack K. Hale and Sjoerd M. Verduyn Lunel, *Introduction to Functional Differential Equations*, Applied Mathematical Sciences **99**, Springer-Verlag, New York, 1993.

[11] R.E. Kalman, On the general theory of control systems, *Proc. First Internl. Cong. IFAC*, Moscow, 1960, Automatic and Remote Control, 1961, pp. 481–92.

[12] R.E. Kalman, Mathematical description of linear dynamical systems, *SIAM J. on Control*, Ser. A, **1**(2), 1963, pp. 152–192.

[13] J.D. Lambert, *Computational Methods in Ordinary Differential Equations*, John Wiley & Sons, New York, 1973.

[14] A. Manitius, G. Payre, R. Roy and H.T. Tran, Computation of eigenvalues associated with functional differential equations, *SIAM J. Sci. Stat. Comput.*, **8**(3), 1987, pp. 222–247.

[15] R.C.H. del Rosario, H.T. Tran and H.T Banks, Proper orthogonal decomposition based control of transverse beam vibrations: Experimental implementation, *IEEE Trans. on Control Systems Technology*, **10**, 2002, pp. 717–726.

[16] W.M. Wonham, On pole assignment in multi-input, controllable linear systems, *IEEE Trans. on AC*, **AC-12**(6), 1967, pp. 660–665.

[17] J. Zabczyk, *Mathematical Control Theory: An Introduction*, Birkhäuser, Boston, 1992.

Chapter 8

Wave Propagation

An area of research in the structural acoustics community which has attracted a great deal of interest in the early nineties is the problem of reducing structure-borne noise levels within an acoustic chamber. A specific example in the aerospace industry was motivated by the development of a class of turboprop engines that are very fuel efficient (see also Section 6.1). These engines, however, produce low-frequency but high amplitude acoustic fields which in turn cause vibrations in the fuselage leading to unwanted interior noise through acoustic/structure interactions. As discussed earlier in Section 6.1, both passive and active control techniques were considered for this problem in frequency domain as well as time domain setting. In addition, mathematical models and approximation techniques were developed for both 2-D and 3-D coupled structural acoustics problem (see, e.g., [1, 3, 4]). In particular, the mathematical models consist of an exterior noise source which is separated from an interior cavity by a thin elastic structure (a beam, plate or shell). The dynamical equations for the shell, plate and beam models under appropriate assumptions are known as Donnell-Mushtari shell, Love-Kirchhoff plate and Euler-Bernoulli beam equations, respectively [5]. In this chapter, the development of the wave equation for the interior acoustic pressure will be considered.

Mathematically we think of sound as perturbations of pressure and density from the "static state" of a fluid. Therefore, we will first consider the working tool of fluid mechanics, the Navier-Stokes equation, which is merely Newton's Second Law of Motion applied to a fluid element.

8.1 Fluid Dynamics

We are all familiar with fluids: ocean waves, air, blood, and so on. The applications of fluid mechanics cover an incredible broad range, from hydraulics, aerodynamics, physical oceanography, atmospheric dynamics, and wind engineering, to cardiovascular medicine and biofluids. Understanding of fluid dynamics has been one of major advances of physics, applied mathematics and engineering over the last hundred years. Indeed modern design of aircraft, spacecraft, automobiles, land and marine structures, to studying efflu-ent discharge into the sea and motions of the atmosphere depend on a clear understanding of the relevant fluid mechanics.

As we shall see, the study of fluids is not so simple, because fluid is extended over space, and when a fluid moves (because of motion of its boundaries) forces are exchanges to the interior of the fluid by the fluid itself. However, underlying all of fluid dynamics are the empirically verified physical principles of mass, momentum, and energy conservations, combined with the laws of thermodynamics. These principles have already been discussed earlier in other applications such as mass transport and heat conduction. We will see in this section how they are applied to obtain the equations of motion for fluid.

We begin by introducing some of the terms and concepts which are often used when describing fluids and fluid motion mathematically.

8.1.1 Newton's Law of Viscosity

A fluid differs from a solid in that a fluid will not come into equilibrium with a shearing force. That is, solids change their shape and deform until a balance is reached between the applied force and internal forces (otherwise the material breaks). Fluids, on the other hand, will deform continuously (that is, flow) under the action of shearing force. One way to imagine a shearing force is to think of a fluid as a stack of thin layers, like a fluid deck of cards. Now, set the cards on a table, rest your hand on top of the deck, then move the hand horizontally. The layers of cards slide over one another, with the top cards being displaced the most, and the bottom card not at all. Unlike a solid, if your hand continues to move horizontally, the top cards will continue moving. This describes how a fluid moves in a non-turbulent flow, one layer sliding over another, although different fluids will oppose its motion by different amounts. This resistance of the fluid to shearing forces is called *viscosity*. One can see this "no-slip" boundary condition by considering the motion of a fluid between two parallel plates of area A, which are separated by a distance Δy (see Figure 8.1). At time $t = 0$ the lower plate moves to right horizontally at a constant velocity Δv_x. As time increases, the fluid velocity profile is as depicted in Figure 8.2. In fact, for large t the fluid velocity distribution is linear along the y-direction (see Figure 8.3). If one measures the force to keep the lower plate moving with constant velocity Δv_x, one finds that it is proportional to the velocity Δv_x and to the area A of the plates, and inversely proportional to the distance Δy. That is,

$$F = -\mu A \frac{\Delta v_x}{\Delta y},$$

where the constant of proportionality μ is called the *viscosity* of the fluid. Now taking the limit as $\Delta y \to 0$, we have

$$F = -\mu A \frac{dv_x}{dy}. \tag{8.1}$$

It is customary to rewrite the above equation in a mere explicit form. We define the shear stress in the x-direction on a fluid surface with normal vector parallel to the y-direction as τ_{yx}, then the equation (8.1) can be rewritten as

$$\tau_{yx} = -\mu \frac{dv_x}{dy}. \tag{8.2}$$

This simply states that the shear stress is proportional to the negative of the velocity gradient, which is the same behavior that one experiences with heat conduction as discussed earlier in Chapter 5 (in which heat flux is proportional to the negative of the temperature gradient). This is known as *Newton's law of viscosity*, and fluids or gases that behave in this fashion are called *Newtonian fluids*. There are, however, quite a few industrially important materials (pastes and highly polymeric materials) which do not obey the relation (8.2), that is, the relation between τ_{yx} and dv_x/dy is not linear. These fluids are referred to as *non-Newtonian fluids*. For non-Newtonian fluids, the viscosity μ is not constant but a function of either shear stress or velocity gradient. The subject of non-Newtonian flow is beyond the scope of this book (we refer the interested reader to [6]).

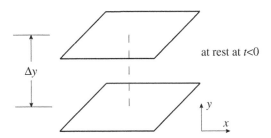

FIGURE 8.1: A fluid initially at rest between two parallel plates.

The shear stress in equation (8.2) can also be interpreted as a *flux of x-directed momentum in the y direction*, which is equivalent to force per unit area. Recall that by flux is meant "rate of flow per unit area." Hence, momentum flux has units of momentum per unit area per unit time. The negative sign in equation (8.2) indicates that momentum tends to go in the direction of decreasing velocity. It should be emphasized that the discussion above is limited to *laminar flow*. That is, at low velocities the fluid flows without lateral mixing, and adjacent layers slide past one another like a deck of playing cards. There are no eddies or swirls of fluids. But not all flow is laminar. At higher velocities fluid swirls erratically. This is called turbulent flow. The onset of turbulent flow depends on the fluid's speed, its viscosity, its density, and the size of the obstacle it encounters. These variables are combined into

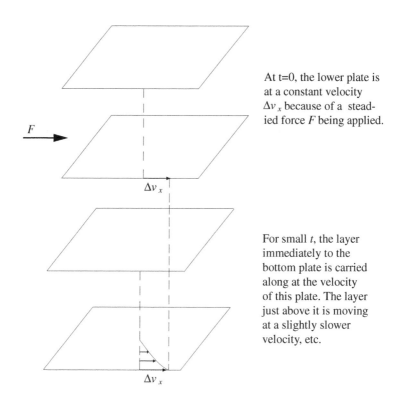

At t=0, the lower plate is at a constant velocity Δv_x because of a steadied force F being applied.

For small t, the layer immediately to the bottom plate is carried along at the velocity of this plate. The layer just above it is moving at a slightly slower velocity, etc.

FIGURE 8.2: Transient velocity profile of a fluid between two parallel plates.

the so-called Reynolds number after the Irish mathematician and physicist Osborne Reynolds (1842-1912). For a flow passing through a straight circular pipe of diameter D, the Reynolds number, which is dimensionless, is defined by

$$\text{Reynolds number} = \frac{\text{density} \times D \times \text{flow speed}}{\text{viscosity}}.$$

For flow in a circular pipe, the flow is laminar when the Reynolds number is less than 2100. When the value is over 4000, the flow is turbulent. In the case where the Reynolds number is between 2100 and 4100, which is called the transition region, the flow can be viscous or turbulent [7].

The viscosity of gases at low density increases with increasing temperature (see, e.g., [6]). In liquids, however, the viscosity usually decreases with in-

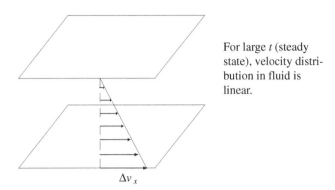

For large t (steady state), velocity distribution in fluid is linear.

Δv_x

FIGURE 8.3: Fluid shear in steady-state between two parallel plates.

creasing temperature. For many applications, viscosity can be ignored and in this case the fluid is called *inviscid*. Another important distinction arises from whether the fluid is *compressible* or *incompressible*. This distinction, which will be made clear mathematically later in the chapter, expresses the elastic properties of the fluid. It describes how the density of the fluid changes in response to changes in pressure and temperature.

In the cgs (centimeter-gram-second) system in which the unit of force is dyne, the unit of viscosity is called a poise or centipoise (cp), in honor of Jean Poiseuille (1797-1869) who developed an improved method for measuring blood pressure. In the international System of Units (SI), the unit of force is newton (N) and the unit of viscosity is newton-second/square meter. One cp is $1 \times 10^{-3} Ns/m^2$. Finally, sometimes the viscosity is also given as the ratio of viscosity to the mass density of the fluid and is called the *kinematic viscosity*. That is, the kinematic viscosity is given by

$$\nu = \mu/\rho,$$

where ρ is the fluid mass density (mass per unit volume). The unit of ν is in m^2/s or cm^2/s. In Table 8.1 some experimental viscosity data are given for some gases and liquids at one atm pressure [6]. Note that at room temperature the viscosity is about one cp for water and about 0.02 cp for air.

TABLE 8.1: Viscosity values of some gases and liquids at atmospheric pressure.

Gases	Temp. ($^{\circ}C$)	μ (cp)	Liquids	Temp. ($^{\circ}C$)	μ (cp)
Air	0	0.01716	Water	0	1.787
Air	20	0.01813	Water	20	1.0019
Air	100	0.02173	Water	100	0.2821
CO_2	20	0.0146	C_2H_5OH	20	1.194
N_2	20	0.0175	Hg	20	1.547
CH_4	20	0.0109	Glycerol	20	1069

8.1.2   Derivative in Fluid Flows

Consider a function $f(t, x, y, z)$ as representing some fluid field variable (such as the velocity, or the pressure, or the density). We want to know how to compute its rate of change with respect to time. Using the chain rule we obtain

$$\frac{df}{dt} = \frac{\partial f}{\partial t} + \frac{\partial f}{\partial x}\frac{dx}{dt} + \frac{\partial f}{\partial y}\frac{dy}{dt} + \frac{\partial f}{\partial z}\frac{dz}{dt}. \tag{8.3}$$

In the above expression the term $\partial f/\partial t$ means the rate of change of f with respect to t at *fixed position* (x, y, z). This is called a *partial derivative* or *Eulerian description*, named after the Swiss mathematician Leonhard Euler (1707-1783).

We can also find the rate of change of f as we follow the fluid or the material. In this case the extended derivative (8.3) must be used and such a form is called the *convective derivative* (or, in some textbooks, the *material derivative* or *Lagrangian description*, named in honor of the French mathematician and mathematical physicist Joseph-Louis Lagrange (1736-1813)). We denote this derivative by

$$\frac{Df}{Dt} = \frac{\partial f}{\partial t} + \frac{\partial f}{\partial x}\frac{dx}{dt} + \frac{\partial f}{\partial y}\frac{dy}{dt} + \frac{\partial f}{\partial z}\frac{dz}{dt}.$$

What this means is that the Lagrangian rate of change in time of any fluid quantity f is made up of two parts: the rate of change in time of f at the instantaneous spatial position of the fluid element and the rate of change of f due to the fact that fluid element is moving from one place to another.

8.1.3   Equations of Fluid Motion

In this section we consider the study of fluid motion as determined by some equations of motion. By this we mean, in the traditional Newtonian sense, an equation which relates the acceleration of the motion of the fluid to the forces that are generating the motion. When we consider the motion of a solid (baseballs, rockets, etc.), it is fairly easy to think of the solid as moving in response to the forces applied to it. On the other hand, fluids are slippery and it is more difficult to think about what it is that is being pushed around. We will get around this obstacle by thinking about an infinitesimal small volume of fluid within the whole body of fluid. In the sequel we will refer to this small volume of fluid as the *fluid element* and it is sufficiently small so that we can consider it as a single point or particle.

The equations which we will derive are known as the Navier-Stokes equations. These equations are coupled with the continuity equation and the equation of state to describe all problems of the viscous flow of a pure isothermal fluid. For nonisothermal fluids, and for multicomponent fluid mixtures, additional equations are needed to describe the conservation of energy and the conservation of individual chemical species. In this section, we will restrict our discussion to isothermal and Newtonian systems.

(a) Continuity Equation

Consider a stationary volume element $\Delta x \Delta y \Delta z$ which is fixed in space as depicted in Figure 8.4. The equation of continuity is developed by considering a mass balance for a fluid flowing through this stationary fluid element:

{rate of mass accumulation} = {rate of mass in} - {rate of mass out}.

Here, it is assumed that we are dealing with fluids which spontaneously generate or destroy material and we are not concerned with any regions with sources or sinks of materials (e.g., taps and plug-holes).

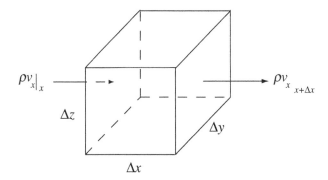

FIGURE 8.4: A fluid element fixed in space through which a fluid is flowing.

We begin by considering the rate of mass entering and leaving the faces perpendicular to the x-axis (see Figure 8.4). Since the product of mass density with the velocity is the mass flux, the rate of mass entering the face at x is $(\rho v_x)|_x \Delta y \Delta z$ and that leaving the face at $x + \Delta x$ is $(\rho v_x)|_{x+\Delta x} \Delta y \Delta z$. Rates of mass entering and leaving the faces perpendicular to the y- and z-

axes can also be derived similarly and are omitted here. The rate of mass accumulation within the fluid element is $(\Delta x \Delta y \Delta z)(\partial \rho / \partial t)$. Substituting all these expressions into the mass balance equation, we obtain

$$\Delta x \Delta y \Delta z \frac{\partial \rho}{\partial t} = \Delta y \Delta z[(\rho v_x)|_x - (\rho v_x)|_{x+\Delta x}]$$
$$+\Delta x \Delta y[(\rho v_z)|_z - (\rho v_z)|_{z+\Delta z}]$$
$$+\Delta x \Delta z[(\rho v_y)|_y - (\rho v_y)|_{y+\Delta y}].$$

Now, dividing both sides by $\Delta x \Delta y \Delta z$ and taking the limits as $\Delta x, \Delta y, \Delta z$ approach zero, we get

$$\frac{\partial \rho}{\partial t} = -\left[\frac{\partial(\rho v_x)}{\partial x} + \frac{\partial(\rho v_y)}{\partial y} + \frac{\partial(\rho v_z)}{\partial z}\right].$$

More conveniently, we may rewrite the above equation in vector form as follows

$$\frac{\partial \rho}{\partial t} = -\nabla \cdot (\rho \vec{v}). \tag{8.4}$$

This is called the *equation of continuity*, which describes the rate of change of density at a fixed point and is precisely the same equation that was derived in Chapter 4 in the context of mass balance (see equation (4.2)). We can convert equation (8.4) into another form by carrying out the actual partial differentiation:

$$\underbrace{\frac{\partial \rho}{\partial t} + v_x \frac{\partial \rho}{\partial x} + v_y \frac{\partial \rho}{\partial y} + v_z \frac{\partial \rho}{\partial z}}_{\frac{D\rho}{Dt}} = \underbrace{-\rho\left[\frac{\partial v_x}{\partial x} + \frac{\partial v_y}{\partial y} + \frac{\partial v_z}{\partial z}\right]}_{-\rho(\nabla \cdot \vec{v})}.$$

Hence, equation (8.4) becomes

$$\frac{D\rho}{Dt} = -\rho(\nabla \cdot \vec{v}), \tag{8.5}$$

where the notation D/Dt denotes the convective derivative as defined earlier. This equation describes the rate of change of density as seen by an observer floating along with the fluid.

In any case, equation (8.4) or (8.5) is simply a statement of *conservation of mass*. Furthermore, these equations can be derived for an arbitrary shape of fluid element instead of a rectangular fluid element as we have done above.

Often in engineering with liquids that are relatively incompressible, the density ρ is essentially constant. In this case,

$$(\nabla \cdot \vec{v}) = 0.$$

This is the *incompressibility condition* and it is applicable to viscous or inviscid fluids. Note that for the above equation to be valid, ρ must remain constant for a fluid element as it moves along a path following the fluid motion (that is, $D\rho/Dt = 0$ employing the Lagrangian description).

(b) Momentum Equations

Fundamental to the derivation of equations of motion is the idea of applying the physical principle of conservation of linear momentum to an arbitrary fluid element, which is of the form:

(rate of momentum accumulation) = (rate of momentum in)

$-$(rate of momentum out)

$+$(sum of forces acting on system).

Similar to the derivation of the continuity equation we will apply the above momentum balance to a stationary fluid element of volume $\Delta x \Delta y \Delta z$ as shown in Figure 8.5. We begin by considering the rates at which the x component of the momentum enters through each of the surfaces. The y and z components of the momentum can be described analogously.

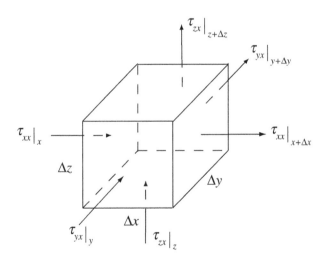

FIGURE 8.5: A fluid element of volume $\Delta x \Delta y \Delta z$ fixed in space through which the x-component of the momentum is transported.

Momentum flows into and out of a fluid element by two mechanisms. One mechanism is by virtue of the bulk fluid flow (convection). The second mechanism is by virtue of the velocity gradient (molecular transfer). This can be seen by considering the interaction between two adjacent layers of a fluid, which have different velocities (and hence different momentum). The random motions of the molecules in the faster moving layer send some of the molecules into the slower moving layer, where they collide with slower moving molecules and thus speed them up (or increase their momentum). Similarly, molecules in the slower moving layer slow down those in the faster moving layer. This exchange of molecules between layers produce a transfer of momentum by virtue of the velocity gradient (from high velocity to low velocity layers).

The rate at which the x component of momentum enters the face at x by convection is

$$((\text{rate of mass in})v_x)|_x = (\rho v_x v_x)|_x \Delta y \Delta z.$$

The rate at which it leaves the face at $x + \Delta x$ is

$$((\text{rate of mass out})v_x)|_x = (\rho v_x v_x)|_{x+\Delta x} \Delta y \Delta z.$$

Note that the product of (ρv_x) with v_x gives the momentum flux (with unit momentum$/s \cdot m^2$). The rate at which the x component of momentum enters the face at y is $(\rho v_y v_x)|_y \Delta x \Delta z$. Similar expressions can be written for the remaining three faces. Combining these momentum fluxes we obtain the following expression for the net convective x momentum flow into the fluid element $\Delta x \Delta y \Delta z$

$$\Delta y \Delta z \left[\rho v_x v_x|_x - \rho v_x v_x|_{x+\Delta x} \right] + \Delta x \Delta z \left[\rho v_y v_x|_y - \rho v_y v_x|_{y+\Delta y} \right]$$
$$+ \Delta x \Delta y \left[\rho v_z v_x|_z - \rho v_z v_x|_{z+\Delta z} \right].$$

Analogously, the net x component of the momentum by molecular transfer is given by

$$\Delta y \Delta z \left[\tau_{xx}|_x - \tau_{xx}|_{x+\Delta x} \right] + \Delta x \Delta z \left[\tau_{yx}|_y - \tau_{yx}|_{y+\Delta y} \right]$$
$$+ \Delta x \Delta y \left[\tau_{zx}|_z - \tau_{zx}|_{z+\Delta z} \right].$$

These products of molecular fluxes of momentum with areas may be considered as friction forces due to shearing. Here, τ_{xx} is the normal stress on the x face, τ_{zx} is the x-directed tangential (or shear) stress on the z face and τ_{yx} is the x-directed shear stress on the y face.

There are two different kinds of forces acting on the fluid element. The first are called *body forces* which act throughout the whole volume of the fluid element (not just on its edges such as surfaces). For fluids in a gravitational field (as on earth where it is effectively constant), this body force is known as the gravitational force. The gravitational force g_x per unit mass in the x direction multiplied by the mass of the element $\rho \Delta x \Delta y \Delta z$ to give the body force

$$\rho g_x \Delta x \Delta y \Delta z.$$

The second kind of forces exist by virtue of the fluid element actually being surrounded by other fluid elements. For a fluid element in the body of a fluid, the rest of the fluid can only exert forces by contact; that is, only at the surface of the fluid element. These forces are called surface forces or pressure forces. The net pressure force acting on the fluid element in the x direction is the difference between the force acting at the face x and that at $x + \Delta x$. This is written as

$$\Delta y \Delta z (p|_x - p|_{x+\Delta x}).$$

Finally, the rate of momentum accumulation in the fluid element of volume $\Delta x \Delta y \Delta z$ is given by

$$\Delta x \Delta y \Delta z \frac{\partial(\rho v_x)}{\partial t}.$$

Substituting the above expressions into the momentum balance equation, we obtain the following expression

$$\begin{aligned}
\Delta x \Delta y \Delta z \frac{\partial(\rho v_x)}{\partial t} = {} & \Delta y \Delta z \left[\rho v_x v_x|_x - \rho v_x v_x|_{x+\Delta x}\right] \\
& + \Delta x \Delta z \left[\rho v_y v_x|_y - \rho v_y v_x|_{y+\Delta y}\right] \\
& + \Delta x \Delta y \left[\rho v_z v_x|_z - \rho v_z v_x|_{z+\Delta z}\right] \\
& + \Delta y \Delta z \left[\tau_{xx}|_x - \tau_{xx}|_{x+\Delta x}\right] \\
& + \Delta x \Delta z \left[\tau_{yx}|_y - \tau_{yx}|_{y+\Delta y}\right] \\
& + \Delta x \Delta y \left[\tau_{zx}|_z - \tau_{zx}|_{z+\Delta z}\right] \\
& + \Delta y \Delta z \left(p|_x - p|_{x+\Delta x}\right) + \rho g_x \Delta x \Delta y \Delta z.
\end{aligned}$$

Dividing by $\Delta x \Delta y \Delta z$ and then taking the limits as $\Delta x, \Delta y, \Delta z$ approach zero, we obtain

$$\begin{aligned}
\frac{\partial}{\partial t}(\rho v_x) = {} & -\left(\frac{\partial}{\partial x}\rho v_x v_x + \frac{\partial}{\partial y}\rho v_y v_x + \frac{\partial}{\partial z}\rho v_z v_x\right) \\
& -\left(\frac{\partial}{\partial x}\tau_{xx} + \frac{\partial}{\partial y}\tau_{yx} + \frac{\partial}{\partial z}\tau_{zx}\right) - \frac{\partial p}{\partial x} + \rho g_x.
\end{aligned}$$

The y and z components of the equation of motion, which can be obtained analogously, are given by

$$\begin{aligned}
\frac{\partial}{\partial t}(\rho v_y) = {} & -\left(\frac{\partial}{\partial x}\rho v_x v_y + \frac{\partial}{\partial y}\rho v_y v_y + \frac{\partial}{\partial z}\rho v_z v_y\right) \\
& -\left(\frac{\partial}{\partial x}\tau_{xy} + \frac{\partial}{\partial y}\tau_{yy} + \frac{\partial}{\partial z}\tau_{zy}\right) - \frac{\partial p}{\partial y} + \rho g_y
\end{aligned}$$

and

$$\begin{aligned}
\frac{\partial}{\partial t}(\rho v_z) = {} & -\left(\frac{\partial}{\partial x}\rho v_x v_z + \frac{\partial}{\partial y}\rho v_y v_z + \frac{\partial}{\partial z}\rho v_z v_z\right) \\
& -\left(\frac{\partial}{\partial x}\tau_{xz} + \frac{\partial}{\partial y}\tau_{yz} + \frac{\partial}{\partial z}\tau_{zz}\right) - \frac{\partial p}{\partial z} + \rho g_z,
\end{aligned}$$

respectively. In vector form, these equations become:

$$\frac{\partial}{\partial t}\rho\vec{v} = -\nabla \cdot \rho\vec{v}\vec{v} - \nabla \cdot \vec{\tau} - \nabla p + \rho\vec{g}, \tag{8.6}$$

where

- $\frac{\partial}{\partial t}\rho\vec{v}$ is the rate of increase of momentum per unit volume;

- $\nabla \cdot \rho\vec{v}\vec{v}$ is the "dyadic product" of $\rho\vec{v}$ and $\vec{v}$ and represents the rate of momentum loss by convection per unit volume. It should be cautioned that $\nabla \cdot \rho\vec{v}\vec{v}$ is not a simple divergence because of the tensorial nature of $\rho\vec{v}\vec{v}$;

- $\nabla \cdot \vec{\tau}$ is the stress tensor term from viscous transfer. Again, this is not a simple divergence;

- ∇p is the pressure force per unit volume;

- $\rho\vec{g}$ is the gravitational force on the fluid element per unit volume.

Using the equation of continuity (8.4) we can rewrite equation (8.6) as

$$\rho\frac{D\vec{v}}{Dt} = -\nabla p - [\nabla \cdot \vec{\tau}] + \rho\vec{g}, \tag{8.7}$$

where

- $\rho\frac{D\vec{v}}{Dt}$ is the mass per unit volume times acceleration,

- ∇p is the pressure force on the fluid element per unit volume,

- $\nabla \cdot \vec{\tau}$ is the viscous force on the fluid element per unit volume,

- $\rho\vec{g}$ is the gravitational force on the fluid element per unit volume.

Hence, in this form the momentum balance equation is simply the following statement

$$\text{mass} \times \text{acceleration} = \sum \text{forces},$$

which implies that the momentum balance is equivalent to Newton's second law of motion.

In order to use these equations to determine the velocity profile of the fluid, we must now express stresses in terms of velocities and fluid properties. For Newtonian fluids, the shear stress components in rectangular coordinates are

$$\tau_{xx} = -2\mu\frac{\partial v_x}{\partial x} + \left(\frac{2}{3}\mu - \kappa\right)(\nabla \cdot \vec{v})$$

$$\tau_{yy} = -2\mu\frac{\partial v_y}{\partial y} + \left(\frac{2}{3}\mu - \kappa\right)(\nabla \cdot \vec{v})$$

$$\tau_{zz} = -2\mu\frac{\partial v_z}{\partial z} + \left(\frac{2}{3}\mu - \kappa\right)(\nabla \cdot \vec{v})$$

$$\tau_{xy} = \tau_{yx} = -\mu\left(\frac{\partial v_x}{\partial y} + \frac{\partial v_y}{\partial x}\right)$$

$$\tau_{yz} = \tau_{zy} = -\mu\left(\frac{\partial v_y}{\partial z} + \frac{\partial v_z}{\partial y}\right)$$

$$\tau_{zx} = \tau_{xz} = -\mu\left(\frac{\partial v_z}{\partial x} + \frac{\partial v_x}{\partial z}\right).$$

Here, κ is called the bulk viscosity and is identically zero for low density monatomic gases and is often neglected in dense gases and liquids. For the remainder of this section, κ is assumed to be zero. These expressions for shear stresses, which are more general statements of Newton's Law of Viscosity, describe a more complex fluid flowing in all directions. In the case where the fluid flows in the x direction between two parallel plates as depicted in Figure 8.1, the above expressions reduce naturally to equation (8.2). Since a detailed derivation of these general expressions for shear stresses is beyond the scope of this book, we refer the interested reader to [9].

Substituting the general shear stress components into the momentum balance equation (8.7), we obtain the general equations of motion for a Newtonian viscous fluid with varying density and viscosity:

$$\rho\frac{Dv_x}{Dt} = -\frac{\partial p}{\partial x} + \frac{\partial}{\partial x}\left[2\mu\frac{\partial v_x}{\partial x} - \frac{2}{3}\mu\left(\nabla \cdot \vec{v}\right)\right]$$

$$+\frac{\partial}{\partial y}\left[\mu\left(\frac{\partial v_x}{\partial y} + \frac{\partial v_y}{\partial x}\right)\right] + \frac{\partial}{\partial z}\left[\mu\left(\frac{\partial v_z}{\partial x} + \frac{\partial v_x}{\partial z}\right)\right] + \rho g_x$$

$$\rho\frac{Dv_y}{Dt} = -\frac{\partial p}{\partial y} + \frac{\partial}{\partial y}\left[2\mu\frac{\partial v_y}{\partial y} - \frac{2}{3}\mu\left(\nabla \cdot \vec{v}\right)\right]$$

$$+\frac{\partial}{\partial x}\left[\mu\left(\frac{\partial v_y}{\partial x} + \frac{\partial v_x}{\partial y}\right)\right] + \frac{\partial}{\partial z}\left[\mu\left(\frac{\partial v_z}{\partial y} + \frac{\partial v_y}{\partial z}\right)\right] + \rho g_y$$

$$\rho\frac{Dv_z}{Dt} = -\frac{\partial p}{\partial z} + \frac{\partial}{\partial z}\left[2\mu\frac{\partial v_z}{\partial z} - \frac{2}{3}\mu\left(\nabla \cdot \vec{v}\right)\right]$$

$$+\frac{\partial}{\partial x}\left[\mu\left(\frac{\partial v_z}{\partial x} + \frac{\partial v_x}{\partial z}\right)\right] + \frac{\partial}{\partial y}\left[\mu\left(\frac{\partial v_z}{\partial y} + \frac{\partial v_y}{\partial z}\right)\right] + \rho g_z.$$

These equations, along with the equation of continuity, as well as the equation of state and boundary and initial conditions, are used to determine pressure, density, and velocity components in a flowing isothermal fluid.

The above equations are seldom used in their general forms. In particular, when the density and viscosity are both constant, the equations are simplified and we obtain the equations of motion for Newtonian inviscid fluids. These equations are the celebrated *Navier-Stokes equations*, which Navier first derived in 1822 and Stokes obtained independently in 1845. In vector form they are given by

$$\rho \frac{D\vec{v}}{Dt} = -\nabla p + \mu \nabla^2 \vec{v} + \rho \vec{g}.$$

When $\nabla \cdot \vec{\tau} = 0$, equation (8.7) reduces to

$$\rho \frac{D\vec{v}}{Dt} = -\nabla p + \rho \vec{g}.$$

This is the famous Euler equation, which is used in flow study where viscosity effect is negligible. It was first derived in 1755 by the Swiss mathematician Leonhard Euler (1707-1783).

(c) Equation of State

So far, we have one equation for conservation of mass and three equations for momentum balance. This gives us a total of four equations for five unknowns v_x, v_y, v_z, ρ, and p. Therefore, we need one more equation for the determination of the pressure, velocities and density of a fluid flow. This equation, called the *equation of state*, describes the relationship between pressure and density. It is an empirical or constitutive type relationship of the form $p = f(\rho)$. For example, in an adiabatic environment we have $p = C\rho^\gamma$ for some constants C and γ.

For the special case of *low density* fluids or gases, pressure and density are proportional

$$p = A\rho.$$

This is basically *Boyle's Law* where the proportional constant A is a linear function of temperature θ, $A = R(\theta - \theta_o)$. Here, θ_o is the reference temperature which is the same for all gases at low densities. Denote

$$T = \theta - \theta_o.$$

If θ is in degree Celsius, $\theta_o = -273.15°C$. Hence, we have

$$T = \theta_{Celsius} + 273.15 =° \text{K}.$$

We then obtain

$$p = RT\rho.$$

This is the equation of state which says that the pressure is proportional to the temperature and inversely proportional to the density. A gas satisfying this equation over an extended range of pressures and temperatures is known as a perfect (or ideal) gas. No real gases obey these laws exactly, but at ordinary temperatures and pressures that are not more than several atmospheres, the ideal gas law gives a very good approximation to within a few percent of the actual values [7]. In the equation of state, R is the gas law constant and is equal to $\frac{\bar{R}}{M}$, where $\bar{R}$ is the universal gas constant ($8314.3 \text{ kg m}^2/\text{kg mol s}^2 K$) and M is the molecular weight.

8.2 Fluid Waves

In this section we will consider the phenomenon of waves. Our interest is not only because of the particular example considered here, which is sound waves, but also because this is a phenomenon which appears in many contexts throughout physics. Sound waves, visible light waves, radio waves, microwaves, electromagnetic waves, water waves, sine waves, cosine waves, earthquake waves, and waves on a string are just a few of the examples of our daily encounters with waves. In addition to waves, there are a variety of phenomenon in our physical world which resemble waves so closely that we can describe such phenomenon as being wavelike. These include the motion of a pendulum and the motion of a mass suspended by a spring (as described in Chapter 2), which can be thought of as wavelike phenomena.

In general, a wave can be described as a disturbance, usually oscillating, which travels through a medium, transporting energy from one location (its source) to another location without the transport (e.g., convection) of a material medium. Each individual particle of the medium is temporarily displaced and then returns to its original equilibrium positioned. There are three types of waves: mechanical waves, electromagnetic waves, and surface waves. Mechanical waves require a material medium (such as air, water, string) to travel. These waves are further divided into three different types. Transverse waves cause particles of the medium to move perpendicular to the direction of the wave. Longitudinal waves are waves in which particles of the medium move in a direction parallel to the direction of the wave (a sound wave is a classic example of a longitudinal wave). Surface waves cause particles of the medium to undergo a circular motion. Surface waves are neither transverse nor longitudinal. The second type of waves are electromagnetic waves that do not require a medium to travel (light, radio). Finally, matter waves are produced by electrons and particles.

8.2.1 Terminology

We begin by reviewing some basic terminologies and concepts that are commonly used to describe waves.

(a) Travelling Wave

Travelling waves are waves that have both spatial and temporal variations. For example, a sinusoidal travelling wave is represented by

$$A\cos(kx - \omega t),$$

where ω is the angular frequency (or just simply called the frequency) of the wave. The angular frequency specifies how the wave oscillates in time. The

SI unit of ω is rad/s. Another related measurement called frequency specifies the number of vibrations per second and is given by

$$f = \frac{\omega}{2\pi}.$$

The SI unit of f is 1/s, which has been given the name of Hertz after the German physicist Heinrich Hertz (1857-1894). By knowing the wave frequency, we can determine how long it takes for a wave to execute one oscillation. This is called the period and is defined by

$$T = \frac{1}{f} = \frac{2\pi}{\omega}.$$

In addition, we can find the wave velocity from the frequency of a wave and its amplitude by the relation

$$u = \lambda f = \frac{\omega}{k}.$$

Finally, it is important to point out that as time increases, the phase $(kx - \omega t)$ of the wave shifts to lower values, so that for a point on the wave to remain fixed, x must also increase (that is, the wave shifts to the right). Thus, the function $A\cos(kx - \omega t)$ represents a wave that is travelling in the direction of increasing x for $\omega > 0$.

(b) Standing Wave

Standing waves, also known as stationary waves, are waves that have spatial variation but no temporal variation. For example, a sinusoidal stationary wave is described by

$$A \cos kx,$$

where A, the amplitude of the wave, is the distance from a crest to where the wave is at equilibrium. The amplitude is used to measure the energy transferred by the wave. The wave number k specifies how the wave oscillates in space. It is related to the wave's wavelength λ, which is the shortest distance between peaks (the highest points) and troughs (the lowest points), by the relation

$$\lambda = \frac{2\pi}{k}.$$

A second type of standing wave is a wave that is formed by the superposition of two travelling waves with the same amplitude but that travel in opposite directions. Here, the standing wave oscillates in time and space but the wave crests do not move. An example of sinusoidal standing wave is given by

$$A\cos kx \cos \omega t = \frac{A}{2}\cos(kx - \omega t) + \frac{A}{2}\cos(kx + \omega t).$$

8.2.2 Sound Waves

Sound is a wave which is created by vibrating objects and propagated through a medium from one location to another. In order to formulate the propagation of sound in space, it is essential to have some kind of understanding of basic mechanisms and phenomena. Fundamentally, the sound wave is transported from one location to another by means of the particle interaction. If the sound wave is moving through the air, then as one air particle is displaced from its equilibrium position, it exerts a push or pull on its nearest neighbors, causing them to be displaced from their equilibrium position. This particle interaction continues throughout the entire medium, with each particle interacting and causing a disturbance of its nearest neighbors. Because a sound wave is a disturbance which is transported through a medium via the mechanism of medium particle interaction, it is characterized as a mechanical wave.

We now seek to mathematically formulate such a process. We begin by noting that sound waves occur in a medium when there are variations in the pressure. In addition, we would like to describe how the medium density changes as it is displaced. Then, of course, the medium particle is displaced and has a velocity, so that we would have to describe the velocity of the medium particles. In summary, the physics of the phenomenon of sound waves involve three features:

(i) The medium particles move and change the density;

(ii) The change in density corresponds to a change in pressure;

(iii) Pressure changes cause medium particle motion.

Let us first consider feature (ii). For a medium (gas, liquid, or solid), the pressure is some function of density. That is,

$$p = f(\rho). \tag{8.8}$$

Also from (ii), even with a change in density, individual fluid elements still conserve their mass, and so we also have the continuity equation, which is

$$\frac{\partial \rho}{\partial t} = -\nabla \cdot (\rho \vec{v}). \tag{8.9}$$

We now consider the third feature, which is the equation of motion produced by pressure changes. Euler's equations for an inviscid flow are still valid here, since they were derived from the rate of change of linear momentum of a fluid element. Hence, when the dynamical effects of gravity can be neglected (as we shall assume), we have

$$\rho \frac{\partial \vec{v}}{\partial t} + \rho (\vec{v} \cdot \nabla) \vec{v} = -\nabla p. \tag{8.10}$$

As discussed earlier, mathematically we think of sound as perturbations of pressure and density from the "static state" of a fluid. Therefore, we begin by considering the description of the static state. We let $\rho_0(r)$ be the static density, p_0 the static pressure and $\vec{v}_0$ the motion of the fluid. Here $r = (x, y, z)$ represents a point in space. For our case we consider $\vec{v}_0 = \vec{0}$, so the fluid is unmoving in the silent (static) case. If this is substituted into Euler's equation (8.10), $\vec{v}_0 = \vec{0}$ implies that p_0 is a constant, since $\nabla p_0 = \vec{0}$. Note that $\rho_0(r)$ is not necessarily a constant, which allows for damping material in the acoustic cavity.

To introduce sound into the system, we use small perturbations of the above quantities, denoted by $\hat{\rho}(t, r)$, $\hat{p}(t, r)$, $\hat{v}(t, r)$, for perturbation of density, pressure and velocity, respectively, of the fluid and define the relationships

$$\rho(t, r) = \rho_0(r) + \hat{\rho}(t, r) \tag{8.11}$$
$$p(t, r) = p_0 + \hat{p}(t, r) \tag{8.12}$$
$$\vec{v}(t, r) = \vec{v}_0(t, r) + \hat{v}(t, x).$$

However, since $\vec{v}_0 = \vec{0}$, then we have $\vec{v} = \hat{v}$, so for our discussion we will disregard the $\hat{v}$, and only use $\vec{v}$ for the perturbations in the velocity. It is noted that in some textbooks, $\rho(t, r)$ may be written as $\rho_0(r)[1 + \delta(t, r)]$ where in our formulation $\hat{\rho}(t, r) = \rho_0(r)\delta(t, r)$.

(a) Assumptions

For the derivation of the wave equations we make the following standing assumptions:

(i) In the absence of sound, the fluid is found in static equilibrium, as previously described by quantities $\rho_0(r)$, $p_0 = c_1$ for some constant c_1 and $\vec{v}_0 = \vec{0}$;

(ii) We are considering a non-viscous fluid;

(iii) The only energy in acoustic motion is mechanical;

(iv) There is zero heat conductivity, which is related to (iii);

(v) The only forces affecting our system are compressive elastic forces.

(b) Linearization

We will do a linearization of the three system equations (8.8), (8.9) and (8.10) for perturbations of the steady, silent case. Our guiding principle is to retain the first order terms in $\vec{v}$ and $\hat{\rho}$ (or δ) to obtain equations for the fluctuations $\hat{\rho}$ and $\hat{p}$ from the static case. We disregard the higher order terms by the assumption that only small perturbations are considered.

8.2.2.1 Euler's Equation

Substituting equations (8.11) and (8.12) into the Euler's equation (8.10) we obtain

$$(\rho_0 + \hat{\rho})\frac{\partial \vec{v}}{\partial t} + (\rho_0 + \hat{\rho})(\vec{v} \cdot \nabla)\vec{v} = -\nabla(p_0 + \hat{p})$$

$$\rho_0 \frac{\partial}{\partial t}\vec{v} + \underline{\hat{\rho}\frac{\partial}{\partial t}\vec{v}} = -\nabla\hat{p}.$$

The underlined term is a higher order term (both $\hat{\rho}$ and $\vec{v} = \hat{v}$ are "small"), so we disregard it. Hence we obtain

$$\rho_0 \frac{\partial \vec{v}}{\partial t} = -\nabla\hat{p}.$$

8.2.2.2 Equation of Continuity

We repeat this process for the continuity equation (8.9) to obtain

$$\frac{\partial}{\partial t}(\rho_0 + \hat{\rho}) + \nabla \cdot ((\rho_0 + \hat{\rho})\vec{v}) = 0$$

$$\frac{\partial \hat{\rho}}{\partial t} + \nabla \cdot (\rho_0 \vec{v}) + \underline{\nabla \cdot (\hat{\rho}\vec{v})} = 0.$$

After disregarding the underlined higher ordered term, the above equation reduces to

$$\frac{\partial \hat{\rho}}{\partial t} + \nabla \cdot (\rho_0 \vec{v}) = 0.$$

8.2.2.3 Equation of State

We use a first-order approximation of our function f, while noting that $p_0 = f(\rho_0)$. Then

$$p = f(\rho) = f(\rho_0 + \hat{\rho}) \approx f(\rho_0) + f'(\rho_0)\hat{\rho}$$
$$= p_0 + f'(\rho_0)\hat{\rho}.$$

Now since $p - p_0 = \hat{p}$, we obtain

$$p - p_0 = f'(\rho_0)\hat{\rho}$$
$$\hat{p} = f'(\rho_0)\hat{\rho},$$

or $\hat{p} = c^2\hat{\rho}$, where $c^2 \equiv f'(\rho_0) = \frac{\partial p}{\partial \rho}\big|_{\rho_0}$, which is the speed of sound in static material.

An associated parameter, the *compressibility* $\mathcal{K} \equiv \frac{1}{c^2 \rho_0}$ is often encountered in the equation $\mathcal{K}\frac{\partial p}{\partial t} = -\nabla \cdot \vec{v}$, which is the equation of continuity when ρ_0 is constant.

In summary, the first order equations for sound are:

- $\rho_0 \dfrac{\partial \vec{v}}{\partial t} = -\nabla \hat{p}$ (Euler);

- $\dfrac{\partial \hat{\rho}}{\partial t} + \nabla \cdot (\rho_0 \vec{v}) = 0$ (Continuity);

- $\hat{p} = c^2 \hat{\rho}$ (State).

8.2.3 Wave Equations

From our three linearized system equations we can derive three wave equations, one of which is the popular $\phi_{tt} = c^2 \Delta \phi$ for the acoustic potential. In general we assume that $c^2 = c^2(r) = f'(\rho_0(r))$.

(a) Wave equation in pressure

Using the linearized state equation $\hat{\rho} = \frac{\hat{p}}{c^2}$ in the linearized continuity equation to obtain

$$\frac{\partial}{\partial t}\left(\frac{\hat{p}}{c^2}\right) = -\nabla \cdot (\rho_0 \vec{v}).$$

Taking $\frac{\partial}{\partial t}$ on both sides of the equation and using the linearized Euler equation yields

$$\frac{1}{c^2}\frac{\partial^2}{\partial t^2}(\hat{p}) = -\nabla \cdot (\rho_0 \frac{\partial \vec{v}}{\partial t})$$
$$= -\nabla \cdot (-\nabla \hat{p})$$
$$= \Delta \hat{p}$$
$$\frac{\partial^2 \hat{p}}{\partial t^2} = c^2 \Delta \hat{p}.$$

Thus the pressure perturbation $\hat{p}$ satisfy the classical *wave equation*.

(b) Wave equation in velocity

We can derive a wave equation in $\vec{v}$, with some added restrictions, which will become evident. We begin by taking $\frac{\partial}{\partial t}$ of the linearized Euler equation to obtain

$$\rho_0 \frac{\partial^2 \vec{v}}{\partial t^2} = -\nabla \hat{p}_t.$$

But from the continuity equation and the state equation we have

$$\hat{p}_t = c^2 \hat{\rho}_t = c^2[-\nabla \cdot (\rho_0 \vec{v})]$$
$$= -c^2 \nabla \cdot (\rho_0 \vec{v}).$$

Therefore,

$$\rho_0 \frac{\partial^2 \vec{v}}{\partial t^2} = -\nabla[-c^2 \nabla \cdot (\rho_0 \vec{v})]$$
$$\rho_0 \frac{\partial^2 \vec{v}}{\partial t^2} = \nabla[c^2 \nabla \cdot (\rho_0 \vec{v})],$$

which is a wave equation in velocity with non-constant $\rho_0(r)$.

If we assume that ρ_0 was a constant, then $c^2 = f'(\rho_0)$ would be constant. In this case,

$$\frac{\partial^2 \vec{v}}{\partial t^2} = c^2 \nabla[\nabla \cdot \vec{v}].$$

We next use a vector identity $\nabla \times (\nabla \times \vec{w}) = \nabla[\nabla \cdot \vec{w}] - \Delta \vec{w}$ (see Appendix B), to obtain

$$\frac{\partial^2 \vec{v}}{\partial t^2} = c^2 \Delta \vec{v} + c^2 \nabla \times (\nabla \times \vec{v}).$$

Thus, if the flow is irrotational ($\nabla \times \vec{v} = 0$), then we obtain the standard wave equation for $\vec{v}$ in the case of constant density irrotational flow:

$$\frac{\partial^2 \vec{v}}{\partial t^2} = c^2 \Delta \vec{v}.$$

(c) Popular wave equation (potential)

To obtain the third wave equation we have to assume constant density and irrotational flow. But how does irrotational flow affect our system? We define $\vec{\omega} = \nabla \times \vec{v}$ as the vorticity, and return to Euler's equation

$$\rho_0 \frac{\partial \vec{v}}{\partial t} = -\nabla \hat{p}.$$

Now take the curl of both sides:

$$\nabla \times \left(\rho_0 \frac{\partial \vec{v}}{\partial t} \right) = -\nabla \times (\nabla \hat{p}) = 0.$$

If ρ_0 is constant, then the above equation yields $\frac{\partial}{\partial t}(\nabla \times \vec{v}) = 0$. Hence, if we assume that $\vec{\omega}|_{t=0} = 0$, which is a no <u>initial</u> vorticity assumption, then we can integrate $\frac{\partial}{\partial t}(\nabla \times \vec{v}) = 0$ to obtain $\nabla \times \vec{v} = 0$, which is irrotational flow. So the assumptions of constant ρ_0 and no initial vorticity imply that $\vec{\omega} = (\nabla \times \vec{v}) = 0$.

Moreover, recall that if curl $\vec{v} = \nabla \times \vec{v} = 0$, then a scalar function ϕ exists such that $\vec{v} = -\nabla \phi$, which is called the velocity potential. It should be noted that ϕ is not unique! Now return to the linearized Euler's equation and recall that we assumed ρ_0 was constant. We have

$$\rho_0 \frac{\partial \vec{v}}{\partial t} = -\nabla \hat{p}$$
$$-\rho_0 \nabla \phi_t = -\nabla \hat{p}$$
$$\nabla(\rho_0 \phi_t - \hat{p}) = 0.$$

This implies that $\rho_0 \phi_t - \hat{p}$ must be constant with respect to the spatial variables. Hence,

$$\rho_0 \phi_t - \hat{p} = -k(t)$$

or

$$\hat{p} = \rho_0 \phi_t + k(t).$$

We claim that without loss of generality we can take $k \equiv 0$. To see this define another potential by

$$\hat{\phi} = \phi + \frac{1}{\rho_0} \int_0^t k(s) \, ds.$$

Then,

$$\nabla \hat{\phi} = \nabla \phi = -\vec{v}$$

still holds while

$$\rho_0 \hat{\phi}_t = \rho_0 \phi_t + k(t) = \hat{p};$$

that is, we can use this potential for the velocity potential. Therefore, without loss of generality,

$$\hat{p} = \rho_0 \phi_t. \tag{8.13}$$

Combining the continuity equation, the state equation and (8.13) with constant ρ_0, we obtain

$$\frac{\partial \hat{\rho}}{\partial t} = -\nabla(\rho_0 \vec{v})$$

$$\frac{\partial}{\partial t}\left(\frac{\hat{p}}{c^2}\right) = +\nabla(\rho_0 \nabla \phi)$$

$$\frac{\partial}{\partial t}\left(\frac{\rho_0 \phi_t}{c^2}\right) = \rho_0 \Delta \phi$$

$$\frac{\rho_0 \phi_{tt}}{c^2} = \rho_0 \Delta \phi$$

$$\phi_{tt} = c^2 \Delta \phi,$$

which is the usual wave equation for the velocity potential.

(d) Summary

We didn't make any assumptions to obtain $\hat{p}_{tt} = c^2 \Delta \hat{p}$, where we have the linearized state equation $\hat{p} = c^2 \hat{\rho}$. With the assumption that ρ_0 is constant, we also have (i) $\frac{\partial^2 \vec{v}}{\partial t^2} = c^2 \Delta \vec{v}$ where c^2 is constant, (ii) there exists a scalar function ϕ such that $\vec{v} = -\nabla \phi$ and $\hat{p} = \rho_0 \phi_t$, which implies that $\phi_{tt} = c^2 \Delta \phi$.

Now consider what happens when ρ_0 is not constant in space. Then we rewrite the linearized Euler's equation

$$\rho_0 \frac{\partial \vec{v}}{\partial t} = -\nabla \hat{p}$$

as

$$\frac{\partial}{\partial t}(\rho_0 \vec{v}) = -\nabla \hat{p},$$

and define "vorticity" as $\vec{\omega} \equiv \nabla \times \rho_0 \vec{v}$. Now we again assume that the initial vorticity is zero ($\vec{\omega}(0) = 0$) and take the curl of the Euler's equation to obtain

$$\nabla \times \frac{\partial}{\partial t}(\rho_0 \vec{v}) = \nabla \times (-\nabla \hat{p})$$

$$\frac{\partial}{\partial t}(\nabla \times \rho_0 \vec{v}) = 0.$$

Hence with $\vec{\omega}(0) = 0$, we have $\nabla \times \rho_0 \vec{v} = 0$. Therefore, there exists a scalar function Φ such that $\rho_0 \vec{v} = -\nabla \Phi$ and $\vec{v} = -\frac{1}{\rho_0}\nabla \Phi$. Now we return to Euler's equation,

$$\frac{\partial}{\partial t}(-\nabla \Phi) = -\nabla \hat{p}$$

$$\nabla(\hat{p} - \Phi_t) = 0.$$

Then by a similar argument as before, we can without loss of generality take $\hat{p} = \Phi_t$.

Finally, we repeat the process which resulted in equation (8.13), starting with the linearized continuity equation, but without the ρ_0 constant assumption. Thus

$$\frac{\partial \hat{\rho}}{\partial t} = -\nabla(\rho_0 \vec{v})$$

$$\frac{\partial}{\partial t}\left(\frac{\hat{p}}{c^2}\right) = -\nabla \cdot (\rho_0 \vec{v})$$

$$\frac{\partial}{\partial t}\left(\frac{\Phi_t}{c^2}\right) = -\nabla \cdot (-\nabla \Phi)$$

$$\Phi_{tt} = c^2 \Delta \Phi.$$

Hence, we obtain the same result for the wave equation without the ρ_0 constant assumption. In this case $c^2 = f'(\rho_0)$ may depend on the spatial variable.

Finally, in the case of plane symmetric pressure waves, i.e., when the physical properties of a wave are constant along the directions tangent to a family of plane surfaces (so that the waves are effectively one-dimensional), the wave equation for the pressure (and similarly for the other two wave equations) becomes

$$\frac{\partial^2 \hat{p}}{\partial t^2} = c^2 \frac{\partial^2 \hat{p}}{\partial x^2}.$$

In this case, the classical D'Alembert solution, in honor of the French mathematician Jean le Rond d'Alembert (1717-1783), to the wave equation is given by [8]

$$\hat{p}(t,x) = F(t - x/c) + G(t + x/c), \tag{8.14}$$

where F and G are arbitrary functions of t and x that are twice continuously differentiable and represent propagating disturbances. It is also necessary to emphasize regarding solution (8.14) that whatever the initial disturbance profiles of F and G, those profiles are maintained during propagation. Thus, sound waves in the one-dimensional problem propagate essentially without distortion.

8.3 Experimental Modeling of the Wave Equation

In specific applications we have to adapt the functions F and G in the D'Alembert solution for the one-dimensional wave equation to the given initial and boundary conditions. In this section, we will describe a cost-effective physical experiment that one can use to study various types of boundary conditions for acoustic wave propagation in a wave duct. This experiment was motivated from an earlier investigation [2] in which the authors considered several types of boundary conditions in the context of time domain models for acoustic waves. They carried out experiments with four different duct terminations (hardwall, free radiation, foam, wedge) to measure the reflection of harmonic waves by the duct terminations over the range of frequency considered. These reflection coefficients are, in turn, used to estimate the parameters in the mathematical models for time domain boundary conditions. The efforts reported there are the first steps in the development of state space/time domain models for use in the control design problems related to acoustic control of noise in a closed cylinder. The ultimate intent was to model the frequency-dependent impedance of a treated aircraft interior such that the time domain interior pressure response to transient excitation may be predicted. In such applications, one has negligible fluid damping of the acoustic pressure fields. Since the major dissipative mechanism entails the partial absorption/partial reflection that occurs at the fluid/wall interface, it is important in the control of the interior acoustic pressure to model this dissipation accurately.

The general hardware needed to set up this experiment is depicted in Figure 8.6. In this experiment a PVC pipe (readily available at any hardware store) is used to study the effects of different boundary conditions on the acoustic response of an enclosed sound field. Sound waves in the pipe are created with the use of a speaker (which can be bought from RadioShack) mounted at one end of the pipe connected to a function generator (we use a four-channel Hewlett-Packard dynamic signal analyzer, model 35670A, see Figure 8.7). The function generator sends an oscillating current signal to the speaker which causes the speaker's diaphragm to vibrate. As the diaphragm moves outward, the air near the speaker is compressed, creating a small volume at relatively high pressure, which propagates away from the speaker. As the diaphragm moves inward, a low pressure area is created which also propagates away from the speaker. The process of compressions and rarefactions continues with a frequency equivalent to the input signal. A higher input frequency implies that the compression/rarefaction cycle occurs more frequently per second. As the sound wave propagates away from the speaker, they are detected and measured by the electnet condenser microphones (we used the Panasonic Omnidirectional Electnet Condenser Microphone Carthridge, model WM-034) mounted at various locations along the acoustic pipe. We used the HP signal analyzer to monitor both the input signal as well as the signal recorded by the microphone. For this configuration, two termination conditions are investigated. The first is a near *hardwall* condition obtained by terminating the PVC pipe with a reinforced aluminum plate. The second tested case in the wave pipe is a *foam* condition. It is hard to anticipate the exact behavior of this type of termination condition. We leave it as a modeling exercise for the reader (see the project description at the end of this chapter).

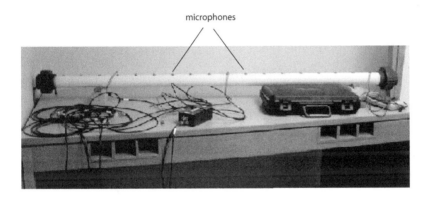

FIGURE 8.6:   Hardware used for studying various types of boundary conditions associated with the one-dimensional wave equation.

FIGURE 8.7:   Hewlett-Packard dynamic signal analyzer.

Project: Sound Wave Propagation in a PVC Pipe

The objective of this project is to study two types of boundary conditions for acoustic waves propagation in a PVC pipe. Experiments with two different boundary conditions (hardwall and foam) are considered using a harmonic oscillator at the other boundary condition. The collected data are then used to obtain reflection coefficients over a wide range of frequencies. The reflection coefficients, in turn, are used to estimate unknown parameters in the models used for boundary conditions.

The acoustic wave motion in a fluid is described by either the acoustic pressure, p, or the velocity potential, ϕ. These two quantities are related by $p(t,x) = \rho\phi_t$, where ρ is the equilibrium density of the fluid (in this case, air at room temperature). In the case where the wavelength of the wave disturbance is large compared to the transverse dimension of the pipe, the wave motion is predominantly parallel to the pipe axis and the sound wave motion is very nearly one-dimensional. That is, the velocity potential ϕ satisfies the following one-dimensional wave equation

$$m\frac{\partial^2 \phi}{\partial t^2} = c^2 \frac{\partial^2 \phi}{\partial x^2}, \qquad 0 < x < l, \tag{8.15}$$

where c is the speed of sound and l is the length of the pipe. Two types of boundary conditions will be considered:

(a) *Oscillating boundaries.* The interaction of the boundary at $x = l$ and the interior pressure is modeled by a damped harmonic oscillator and is described by

$$\delta_{tt} + d\delta_t + k\delta = -\rho\phi_t(t,l). \tag{8.16}$$

Here, δ is the normal displacement of the boundary in the direction interior to the fluid. The coefficients m, d and k are the effective mass, resistance, and the stiffness per unit area of the boundary surface and are assumed to be unknown. In addition, it is also assumed that the boundary surface is not penetrable by the fluid, that is,

$$\delta_t(t) = \phi_x(t,l). \tag{8.17}$$

Recall that the D'Alembert solution to the wave equation (8.15) has the form

$$\phi(t,x) = F(t - x/c) + G(t + x/c), \tag{8.18}$$

where the first term on the right side of (8.18) describes a wave propagating to the right and the second term corresponds to a left propagating wave. From equation (8.17) and by integrating, we obtain

$$\delta(t) = -\frac{1}{c}(\tilde{F}(t) - \tilde{G}(t)), \tag{8.19}$$

where, without loss of generality, the constant of integration is set to zero and $\tilde{F}(t) = F(t - l/c)$ and $\tilde{G}(t) = G(t + l/c)$. Substituting the expression of the solution (8.19) into (8.16) yields

$$m\tilde{G}_{tt} + (d + \rho c)\tilde{G}_t + k\tilde{G} = m\tilde{F}_{tt} + (d - \rho c)\tilde{F}_t + k\tilde{F}. \tag{8.20}$$

Now, assume that the incident wave $\tilde{F}$ to the boundary at $x = l$ (which is generated by a harmonic input at $x = 0$) is a simple harmonic of frequency $\omega/2\pi$, where ω is the angular frequency. That is,

$$\tilde{F}(t) = A_0 e^{i\omega t}, \tag{8.21}$$

so that the right side of (8.20) is a harmonic forcing function. It follows that the *steady state* solution of (8.20) is also harmonic with the same frequency

$$\tilde{G}(t) = R(\omega)A_0 e^{i\omega t}, \tag{8.22}$$

where the complex coefficient, $R(\omega)$, is called the *reflection coefficient*. Substituting equations (8.21) and (8.22) into (8.20) we have a relation for the reflection coefficient for the oscillating boundary condition model given by

$$R(\omega) = \frac{m\omega^2 - i(d - \rho c)\omega - k}{m\omega^2 - i(d + \rho c)\omega - k}. \tag{8.23}$$

(b)*Damped elastic boundaries.* For $d = 0$, $k = 0$ the models (8.16), (8.17) together with the relation $p = \rho\phi_t$ results in the boundary condition

$$mp_x(t, l) + \rho p(t, l) = 0,$$

for the acoustic pressure. This is called a Robin or elastic boundary condition. To include dissipation it is extended by adding a damping term p_t that gives the following boundary condition in terms of the acoustic pressure

$$\alpha p(t, l) + \beta p_t(t, l) + cp_x(t, l) = 0.$$

Assuming harmonic incident wave as previously, show that the damped elastic reflection coefficient is given by the following expression:

$$R(\omega) = \frac{i\omega(1 - \beta) - \alpha}{i\omega(1 + \beta) + \alpha}. \tag{8.24}$$

This project involves the following steps:

1. The acoustic pressure anywhere in the pipe for planar wave propagation is given by the following equation:

$$p(t, x) = A(\omega)e^{i\omega(t - x/c)} + A(\omega)R(\omega)e^{i\omega(t + x/c)}.$$

By measuring the pressure, $p(t, x)$, at a number of axial locations, x_j, and for a specific angular frequency $\tilde{\omega}$, an inverse least squares problem can be formulated to estimate both complex coefficients, $A(\tilde{\omega})$ and $R(\tilde{\omega})$. Considering both physical hardwall and foam type of boundary conditions at $x = l$ and collecting two corresponding sets of experimental data, one can use these to estimate the reflection coefficient $R(\tilde{\omega})$. This data will be denoted by $R^d(\tilde{\omega})$, over the range of frequencies from 100 Hz to 500 Hz.

2. In this problem, we will evaluate how well the oscillating boundary and damped elastic boundary models described by formulas (8.23) and (8.24) fit the experimental data $R^d(\tilde{\omega}_j)$. One approach is to determine the set of parameters, (m, d, k, ρ) and (α, β), so that the functional

$$\sum_{j=1}^{N} |R^d(\tilde{\omega}_j) - R(\tilde{\omega}_j)|^2$$

is minimized. Here, N is the number of measurements R^d at frequencies $f_j = \omega_j / 2\pi$. In your report, discuss which model (8.23) or (8.24), or both, is (or are) best to describe the hardwall and the foam type of boundary conditions.

References

[1] H.T. Banks, W. Fang, R.J. Silcox and R.C. Smith, Approximation methods for control of acoustic/structure models with piezoceramic actuators, *Journal of Intelligent Material Systems and Structures*, **4**(1), 1993, pp. 98–116.

[2] H.T. Banks, G. Propst and R.J. Silcox, A comparison of time domain boundary conditions for acoustic waves in wave guides, *Quarterly of Applied Mathematics*, **LIV**(2), 1996, pp. 249–265.

[3] H.T. Banks, R.J. Silcox and R.C. Smith, The modeling and control of acoustic/structure interaction problems via piezoceramic actuators: 2-D numerical examples, *ASME Journal of Vibration and Acoustics*, **116**(3), 1994, pp. 386–396.

[4] H.T. Banks and R.C. Smith, Modeling and approximation of a coupled 3-D structural acoustics problem, in *Progress in Systems and Control Theory*, K.L. Bowers and J. Lund, eds., Birkhäuser, Boston, 1993, pp. 29–48.

[5] H.T. Banks, R.C. Smith and Y. Wang, *Smart Material Structures: Modeling, Estimation and Control*, John Wiley & Sons, Inc., 1996.

[6] R.B. Bird, W.E. Stewart and E.N. Lightfoot, *Transport Phenomena*, John Wiley & Sons, Inc., New York, 1960.

[7] C.J. Geankoplis, *Transport Processes and Unit Operations*, Prentice-Hall, Inc., 1993.

[8] K.F. Graff, *Wave Motion in Elastic Solids*, Dover Publications, Inc., 1991.

[9] H. Schlichting, *Boundary-layer Theory*, MacGraw-Hill, New York, 1979.

Chapter 9

Size-Structured Population Models

9.1   Introduction: A Motivating Application

The mosquitofish, *Gambusia affinis*, is used throughout the world to control mosquito populations. Indigenous to the southeastern United States and northeastern Mexico, it is one of the most widely distributed of all freshwater fish. When introduced into a rice field, the mosquitofish eat the water-borne mosquito larvae. Consequently, it is thought to be the most widely disseminated natural predator as well as the most popular form of mosquito control.

In spite of their widespread use, the mechanisms underlying the growth of Gambusia populations (and consequently, mosquito control) are not well understood. For example, studies have shown that application of Gambusia early in the rice season leads to fewer mosquito larvae on the average over several fields. However, there is considerable variability among rice fields, with some unstocked fields having fewer larvae than stocked fields.

In the early 1980s a research group from UC-Davis [12] carried out experiments to better understand how Gambusia populations develop in rice fields. Their goal was to achieve better mosquito management through more detailed knowledge of Gambusia population and predation dynamics. Even though the economic implications were substantial, no one really knew how many mosquitofish should be used to stock a rice paddy field. In addition, stocking methods do significantly differ, raising many questions. For example, should all the mosquitofish be added initially, or should they be introduced into the rice paddy field periodically or by some other time dependent schedule?

There are a number of avenues that can be taken to investigate these questions. A control theorist might try to use a general system of ordinary differential equations such as

$$\dot{x} = Ax + Bu$$

and choose a control u (stocking rate perhaps) to improve system behavior (see Chapter 7 for an introduction to the control theory). However, this requires knowing the matrices A and B. At one time control theorists thought biologists might be able to provide A and B, but they unfortunately were not able to do this with any degree of certainty.

Another avenue is to perform many experiments in hope of finding some empirical relationship. The approach that we pursue here is to adapt some sort of reasonable mathematical model to understand the basic dynamics of growth and decline in the mosquitofish population. Several types of population models have been developed over the years to model population dynamics. These include single species models, logistic models, predator/prey models and structured models, each of which will be discussed in the following sections.

9.2 A Single Species Model (Malthusian Law)

The simplest population models are the single species models. Let $p(t)$ denote the population (number) of a given species at time t. Assuming that this population is isolated (that is, there is no net immigration nor emigration), then the rate of change of the population is simply the difference between the birth rate and the death rate

$$\frac{dp}{dt} = \text{birth rate} - \text{death rate}.$$

We further assume that the more individuals there are, the more births and deaths that occur. That is, both the birth rate and death rate are proportional to the number of individuals in the population. Consequently, the birth rate is given by βp and death or mortality rate is μp. In this case, the model becomes

$$\frac{dp}{dt} = \beta p - \mu p$$
$$= \alpha p, \tag{9.1}$$

where $\alpha = \beta - \mu$ represents the net rate of birth/death per individual in the population. Equation (9.1) is a linear first order differential equation and is known as the *Malthusian* law of population growth. If the population of a given species is p_0 at time $t = t_0$, then the solution to the initial-value problem has the form $p(t) = e^{\alpha(t-t_0)}p_0$. Depending on the value of α the solution $p(t)$ will have one of the following three characteristics: (i) when $\alpha > 0$ (more births than deaths) the population will grow exponentially with time, (ii) when α is negative the population will die out, and (iii) when α is equal to zero the population will remain constant and is equal to the initial number of individuals p_0 (see Figure 9.1).

The single species model is so simple that it predicts population outcomes that are clearly unreasonable. Note that the deaths in this model are from "natural causes" or old age. There is no predatory or otherwise harmful

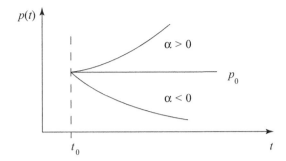

FIGURE 9.1: Graphs of the population $p(t)$.

activities represented in this model. Moreover, when the number of individuals p becomes very large, the single species model cannot be very accurate, since it does not reflect the fact that individual members are now competing with each other for limited living space, natural resources, and food.

9.3 The Logistic Model

Clearly overcrowding will reduce the amount of food, as well as tax other resources such as oxygen levels, etc. In the single species model we can add a crowding term, which will result in more deaths with higher numbers of individuals. A simple first assumption might be that the death rate per individual μ, is a function of the population p. That is, we might take $\mu = \mu(p)$. The simplest form of such a function is linear $\mu(p) = \mu p$, so that the model becomes

$$\frac{dp}{dt} = \beta p - (\mu p)p. \tag{9.2}$$

This equation was first introduced by the Dutch mathematical biologist Verhulst in 1837 and has subsequently become known as the *logistic* equation. The term μp in equation (9.2) simply translates to more deaths occurring when p is large; this is the competition or crowding term.

We observe that if p is small, $-\mu p^2$ is negligible and the model reduces to the Malthusian law. On the other hand, if p is large, $-\mu p^2$ serves to slow down the rapid rate of increase. In either case, for $\mu \neq 0$, the equation is readily solved analytically via standard techniques. Using the method of separation of variables, we rewrite the differential equation

$$\frac{dp}{dt} = \beta p - \mu p^2, \qquad p(t_0) = p_0$$

as

$$\frac{dp}{\beta p - \mu p^2} = dt.$$

Hence we find

$$p(t) = \frac{\beta p_0}{\mu p_0 + (\beta - \mu p_0)e^{-\beta(t-t_0)}},$$

the graph of which is depicted in Figure 9.2. This solution is often written as

$$p(t) = \frac{K p_0}{p_0 + (K - p_0)e^{-\beta(t-t_0)}}$$

corresponding to the equation being written as

$$\frac{dp}{dt} = \beta p \left(1 - \frac{p}{K}\right),$$

where $K = \beta/\mu$ is the population's *carrying capacity* and β is called the *intrinsic growth rate*.

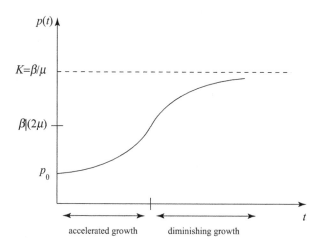

FIGURE 9.2:   Graph of the solution to the logistic model.

We remark that regardless of the initial population p_0, the number of individuals always approaches the limiting value $K = \beta/\mu$ as $t \to \infty$. Furthermore, since

$$\frac{d^2p}{dt^2} = \frac{d}{dt}\left(\frac{dp}{dt}\right)$$
$$= \frac{d}{dt}(\beta p - \mu p^2)$$
$$= \beta\frac{dp}{dt} - 2\mu p\frac{dp}{dt}$$
$$= (\beta - 2\mu p)(\beta - \mu p)p,$$

it follows that if $p < \frac{\beta}{2\mu}$, then $\frac{d^2p}{dt^2} > 0$ and p is thus concave up. On the other hand, if $p > \frac{\beta}{2\mu}$ (and $p < \frac{\beta}{\mu}$), then $\frac{d^2p}{dt^2} < 0$ and p is concave down. Hence, the graph of p has the form as depicted in Figure 9.2. Such a curve is called a *logistic*, or *S-shaped* curve. From its shape, the time period before the population reaches $\frac{\beta}{2\mu}$ is known as the period of accelerated growth. After this period, the rate of growth decreases and asymptotically reaches zero. This is a period of diminishing growth.

The logistic model is sometimes also called the Verhulst-Pearl model (it was developed by Verhulst [21] and later rediscovered and popularized by Pearl [19]). It has been widely used [17] for many years in certain applications. Its primary feature, the population saturation, is biologically realistic if nothing else is preying on the population. However, this model is not adequate in a predator/prey situation.

9.4   A Predator/Prey Model

In the mid-1920s, Italian biologist Umberto d'Aucona studied the percent of total catch of selachians (a group of fish comprising the sharks, skates, and rays) in the Mediterranean port of Port Fiume, Italy. The data is tabulated in Table 9.1 for the period from 1914 to 1923 [13].

He was puzzled by the very large increase of selachians during World War I (1914-1918). He reasoned that selachians increased due to the reduced level of fishing during the war. Therefore, there were more fish available as food for the selachians, and hence the selachian population multiplied. However, this explanation was not satisfactory since one did not have more food fish (supposedly to be eaten by sharks) during this period.

After exhausting all biological explanations, in 1926 [22] he turned to his colleague, the famous Italian mathematician Vito Volterra, for help. Volterra formulated a mathematical model for the growth of selachians and their prey, food fish, by separating all food fish into the prey population and selachians into the predator population.

TABLE 9.1:   Percent of total catch of selachians.

1914	1915	1916	1917	1918
11.9%	21.4%	22.1%	21.2%	36.4%
1919	1920	1921	1922	1923
27.3%	16.0%	15.9%	14.8%	10.7%

Let the number of predators and prey at time t be $N(t)$ and $E(t)$ (the edibles), respectively. A simple assumption is that the population of edibles will grow exponentially without the predators. In addition, the prey death rate depends on both E and N (since they are eaten by predators). Similarly, since the predators need the edibles to live, their birth rate will be dependent on E and N as well. Finally, with no edibles, the predators are assumed to die out exponentially. Then, we can write the following system of differential equations for the predator/prey model:

$$\frac{dN}{dt} = (\beta_N E)N - \mu_N N,$$
$$\frac{dE}{dt} = \beta_E E - (\mu_E N)E. \tag{9.3}$$

The system of equations (9.3), which is also called the Lotka-Volterra model, has two equilibrium solutions:

$$N^e = E^e = 0$$

and

$$N^e = \frac{\beta_E}{\mu_E}, \qquad E^e = \frac{\mu_N}{\beta_N}.$$

Moreover, it has the following families of solutions:

(i) $E(t) = E_0 e^{\beta_E t}, \; N(t) = 0,$
(ii) $N(t) = N_0 e^{-\mu_N t}, \; E(t) = 0.$

Hence, both the E and N axes are orbits of (9.3). This implies that every solution E and N of (9.3) that starts in the first quadrant, $E > 0$ and $N > 0$, will remain there for all $t \geq t_0$ (which is guaranteed by the uniqueness result of the solution to (9.3)). Furthermore, the orbits for $E, N \neq 0$ can be found by solving the following equation

$$\frac{dN}{dE} = \frac{-\mu_N N + \beta_N EN}{\beta_E E - \mu_E NE},$$

which, after one separates variables and integrates both sides, yields

$$\frac{N^{\beta_E}}{e^{\mu_E N}} \frac{E^{\mu_N}}{e^{\beta_N E}} = k_1. \tag{9.4}$$

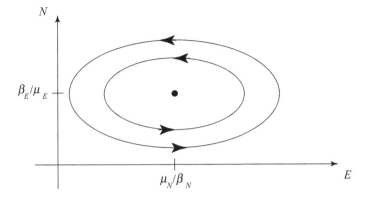

FIGURE 9.3: Orbital solutions of the predator/prey model.

Equation (9.4) defines a family of closed curves for $E, N > 0$ which are depicted in Figure 9.3.

As shown in Figure 9.3, the solutions to the predator/prey model are periodic functions. The Lotka-Volterra model forms the basis of many models used today in the analysis of population dynamics. However, in its original form it has some significant problems. First, neither equilibrium point is stable (see, e.g., Figure 9.3). In addition, many ecologists/biologists refused to accept Volterra's model. They cited the experiments of G.F. Gause (1934) with two species of protozoan (one of which feeds on the other). In all experiments, the predators, *Didinium*, quickly destroyed the prey, *Paramecium*, and then died of starvation. In this case, the number of individuals of both species decays to zero and clearly does not oscillate indefinitely. Obviously the Volterra model does not take into account that bigger fish eat more and that the size varies greatly in the population. Therefore, an approach to introduce these factors into a model is to consider size-structured modeling.

9.5   A Size-Structured Population Model

The logistic and Lotka-Volterra models are both **aggregate** models. That is, they assume that all individuals are identical in characteristics and behavior (fish are all of the same size, for example). Gause's predator/prey experiments indicate that this assumption is not very realistic.

We can attempt to model the individuals or members of a system by behavior or characteristic. This might produce a more realistic model than the aggregate model but it is also much more complicated. In 1967 Sinko and Streifer [20] balanced this trade-off by letting all individuals share some common traits, but permitted variation in size. Their formulation and its generalizations have subsequently been used widely in biological modeling [18].

Let $u(t, x)$ be the number of individuals of size x at time t. If we assume that the species has M distinct size classes $x_1, x_2, ..., x_M$, the total population $N(t)$ at time t will be given by

$$N(t) = \sum_{i=1}^{M} u(t, x_i).$$

This is *size discrete modeling*. Here, growth is a jump from one size class to the next. For growth to be continuous, we will let $x = x(t)$ be a continuous function of t. Now we cannot determine how many individuals are in a specific size class, but instead we calculate the number in an interval of size. We use $u(t, x)$ to denote *size density* (in numbers per unit size) and calculate the number of individuals between size a and b at time t by

$$N_{ab}(t) = \int_a^b u(t, \xi)d\xi.$$

It is important to note that x is *not* a spatial variable and has nothing to do with the location of the individual in the medium. It actually denotes *size*. Since $x(t)$ is size, the flux of $x(t)$ is defined in terms of growth from size x to $x+\Delta x$. Also the size density term, $u(t, x)$, would have units of *individuals/size* that is very different from a location density data, which might have units of *individuals/length^3*.

As already mentioned above, to balance an aggregate model and individual model, Sinko and Streifer grouped individuals sharing common traits together. Specifically, they make the following assumptions:

1. The growth rate, $g > 0$, of same sized individuals is the same. That is,

$$\frac{dx}{dt} = g(t, x).$$

   The simplifying effect of this equation for growth is that the growth of all sizes of individuals is governed by this one equation. Moreover, it is assumed that g is a continuous function.

2. Individuals of the same size have the same likelihood of death. In a simple version, all sizes will have the same death rate. This gives the following basic equation for "simple" death:

$$\frac{dN}{dt} = -\mu N(t),$$

   where μ is the constant of proportionality of mortality. A more complicated model will have $\mu = \mu(x)$ so that mortality is a function of size (i.e., a large individual might be more likely to die than a small individual).

3. The population is sufficiently large to be treated with a continuum model.

4. There is a "smallest" and a "largest" size ($x_0 \leq x \leq x_1$).

5. Birth (also called recruitment) rate is proportional to the population size density and is given by

$$R(t) = \int_{x_0}^{x_1} k(t, \xi) u(t, \xi) \, d\xi,$$

where $k(t, \xi)$ is the size-dependent fertility term also called the fecundity function.

We begin the model derivation by considering first the simple case. Here, we assume that there are no births and no deaths; thus, the population size is constant. That is,

$$\int_{x_0}^{x_1} u(t, x) \, dx = C,$$

where the constant C is the total population. Now, we consider the population from size a to b at time t_0. Then,

$$N_{ab}(t_0) = \int_a^b u(t_0, \xi) \, d\xi,$$

where $N_{ab}(t_0)$ is the shaded area depicted in Figure 9.4. Next, we consider

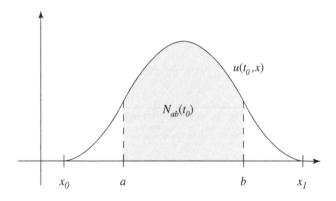

FIGURE 9.4: Total population from size a to b at time t_0.

this same distribution at some later time, say t_1, where $t_1 > t_0$. In that time the fish that were size a grow by Δa and the size b fish grow by Δb. However, the number of fish between sizes $a + \Delta a$ and $b + \Delta b$ should be the same as the number of fish between sizes a and b. That is,

$$\int_a^b u(t_0, \xi)\, d\xi = \int_{a+\Delta a}^{b+\Delta b} u(t_1, \xi)\, d\xi.$$

This is certainly not true if fish from other size classes are entering this size interval. We can easily show and therefore subsequently assume that this cannot happen.

Let $t = t_0$ and assume that we have $x^{(1)}(t_0) < x^{(2)}(t_0)$ for two classes $x^{(1)}$, $x^{(2)}$. We will now show that $x^{(1)}(t) < x^{(2)}(t)$ for all $t > t_0$. Considering the simple growth functions with initial functions

$$\dot{x}^{(1)} = g(t, x^{(1)}), \qquad x^{(1)}(t_0) = x_0^1,$$
$$\dot{x}^{(2)} = g(t, x^{(2)}), \qquad x^{(2)}(t_0) = x_0^2,$$

we will prove our assertion by contradiction. Assuming that $x_0^1 < x_0^2$ and $x^{(1)}(t) > x^{(2)}(t)$ for some $t > t_0$ and that the growth function g is continuous, we find at some time $t_{new} > t_0$ that the two sizes $x^{(1)}$ and $x^{(2)}$ are the same (see Figure 9.5). However, they grow at different rates from t_{new} to t_{later} which is a contradiction to our assumption that individuals of the same size grow at the same rate. This is indeed a consequence of uniqueness of solution to the ordinary differential equation for $x(t)$.

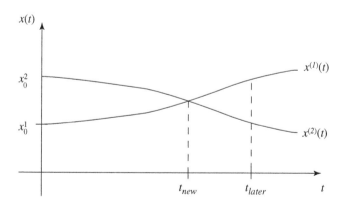

FIGURE 9.5: Size trajectories.

Now let $a, b \in (x_0, x_1)$, where $a < b$, and let $x(t; t_0, \eta)$ denote the unique solution to

$$\dot{x} = g(t, x)$$
$$x(t_0) = \eta.$$

Since there are no births nor deaths and individuals cannot "jump" into a different size interval, we have

$$\int_a^b u(t_0, \xi)\, d\xi = \int_{x(t;t_0,a)}^{x(t;t_0,b)} u(t, \xi)\, d\xi. \tag{9.5}$$

To obtain the differential version of the conservation formula (9.5) we differentiate both sides of the equation with respect to t to obtain

$$\frac{d}{dt}\int_a^b u(t_0, \xi)\, d\xi = \frac{d}{dt}\int_{x(t;t_0,a)}^{x(t;t_0,b)} u(t, \xi)\, d\xi.$$

Obviously the left side is zero. Using Leibnitz's rule [16] on the right side we have

$$0 = \int_{x(t;t_0,a)}^{x(t;t_0,b)} \frac{\partial}{\partial t} u(t, \xi)\, d\xi + u(t, x(t;t_0,b))\dot{x}(t;t_0,b) - u(t, x(t;t_0,a))\dot{x}(t;t_0,a)$$

$$= \int_{x(t;t_0,a)}^{x(t;t_0,b)} \frac{\partial}{\partial t} u(t, \xi)\, d\xi + u(t, x(t;t_0,b))g(t, x(t;t_0,b))$$

$$-u(t, x(t;t_0,a))g(t, x(t;t_0,a))$$

$$= \int_{x(t;t_0,a)}^{x(t;t_0,b)} \frac{\partial}{\partial t} u(t, \xi)\, d\xi + \int_{x(t;t_0,a)}^{x(t;t_0,b)} \frac{\partial}{\partial \xi}(u(t, \xi)g(t, \xi))\, d\xi,$$

$$= \int_{x(t;t_0,a)}^{x(t;t_0,b)} \left\{ \frac{\partial}{\partial t} u(t, x) + \frac{\partial}{\partial \xi}(u(t, \xi)g(t, \xi)) \right\}\, d\xi.$$

Since $(x(t, t_0; a), x(t, t_0; b))$ is an arbitrary interval of sizes, the integrand must be zero and we obtain the equation of conservation

$$\frac{\partial}{\partial t} u(t, x) + \frac{\partial}{\partial \xi}(u(t, \xi)g(t, \xi)) = 0. \tag{9.6}$$

We now present another way to derive the conservation equation (9.6) by considering flux balancing. That is,

$$\text{rate of change of population in the size interval } (a, b) =$$
$$\text{rate of individuals entering } (a, b)$$
$$-\text{rate of individuals leaving } (a, b).$$

Let the interval be $[x, x + \Delta x]$, then we have

$$\frac{d}{dt} N_{x,x+\Delta x}(t) = g(t, x)u(t, x) - g(t, x + \Delta x)u(t, x + \Delta x) \tag{9.7}$$

$$\frac{d}{dt}\int_x^{x+\Delta x} u(t, \xi)\, d\xi = g(t, x)u(t, x) - g(t, x + \Delta x)u(t, x + \Delta x)$$

$$\frac{d}{dt}\frac{1}{\Delta x}\int_x^{x+\Delta x} u(t, \xi)\, d\xi = \frac{g(t, x)u(t, x) - g(t, x + \Delta x)u(t, x + \Delta x)}{\Delta x}.$$

Now taking the limit as Δx approaches zero, we obtain

$$\frac{d}{dt}u(t,x) = -\frac{\partial}{\partial x}(g(t,x), u(t,x)),$$

or

$$\frac{d}{dt}u(t,x) + \frac{\partial}{\partial x}(g(t,x), u(t,x)) = 0.$$

This equation is a hyperbolic partial differential equation. Hence the growth follows the "characteristics" (in this case, solutions $(t, x(t))$ of the equation $\frac{dx}{dt} = g(t,x)$). Let us consider what happens to the individuals that start at size x_0 at time t_0. At some later time $t_1 > t_0$, all of the individuals in our system have size $x(t_1) > x(t_0; t_0, x_0)$. In addition, since x_1 is the largest size, $x(t) < x_1$ for all t. Therefore, as t becomes large, all individuals will be at a size close to x_1, the maximum size; that is, the population is "bunching up" near x_1 and there are no individuals in the shaded region depicted in Figure 9.6. This is a major drawback to conservation. To overcome this undesirable characteristic we now consider adding births and deaths.

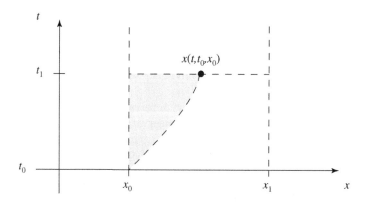

FIGURE 9.6:   Growth characteristic of the conservation equation.

Recalling the flux balance equation (9.7), we now add a death rate term to the right side:

$$\frac{\partial}{\partial t}\int_x^{x+\Delta x} u(t,\xi)\, d\xi = g(t,x)u(t,x) - g(t,x+\Delta x)u(t,x+\Delta x)$$

$$-\text{death rate term.} \tag{9.8}$$

The death term depends on a mortality factor $\mu(t,x)$, as well as $u(t,x)$ and

Δx. Therefore, the flux balance equation (9.8) becomes

$$\frac{\partial}{\partial t} \int_x^{x+\Delta x} u(t,\xi)\, d\xi = g(t,x)u(t,x) - g(t,x+\Delta x)u(t,x+\Delta x)$$
$$- \int_x^{x+\Delta x} \mu(t,\xi)u(t,\xi)\, d\xi. \tag{9.9}$$

Now dividing both sides of equation (9.9) by Δx and letting Δx go to zero we obtain

$$\frac{\partial}{\partial t} u(t,x) = -\frac{\partial}{\partial x}(g(t,x)u(t,x)) - \mu(t,x)u(t,x). \tag{9.10}$$

Equation (9.10) is known as the McKendrick-Von Foerster equation or the Sinko-Streifer equation [20]. The functions $g(t,x)$ and $\mu(t,x)$ correspond respectively to the growth rate of an individual of size x at time t and the fraction of individuals of size x dying at time t. To complete the description of this mathematical model requires the specification of an initial condition

$$u(0,x) = \Phi(x)$$

and a boundary condition. We assume that all births entering the population begin at the smallest size x_0, for simplicity. More specifically, we have:

$$\text{rate of population entering at } x_0 = \text{birth rate}$$
$$g(t,x_0)u(t,x_0) = \int_{x_0}^{x_1} k(t,\xi)u(t,\xi)\, d\xi,$$

or

$$R(t) = g(t,x)u(t,x)|_{x=x_0} = \int_{x_0}^{x_1} k(t,\xi)u(t,\xi)\, d\xi. \tag{9.11}$$

Here R is known as the recruitment rate. We note that when the newborns enter the system, they follow the characteristic growth curves just like other individuals. The addition of (9.11) essentially completes the specification of the mathematical model (9.10). However, since x_1 is the maximum attainable size, we also impose the physical condition

$$g(t,x)u(t,x)|_{x=x_1} = 0.$$

If the functions $g(t,x)$, $\mu(t,x)$, $R(t)$ and $\Phi(x)$ are known explicitly, this system can be solved using the *method of characteristics* [14]. To explain this, we begin by considering the simpler case where the growth rate function is assumed to be constant, $g(t,x) = a$, and the mortality factor $\mu(t,x) = 0$. That is,

$$u_t + au_x = 0,$$
$$u(0,x) = \Phi(x).$$

Note that the total derivative or directional derivative of u is given by

$$du = u_t dt + u_x dx,$$
$$= \left(u_t + u_x \frac{dx}{dt} \right) dt.$$

Hence, in the direction of $\frac{dx}{dt} = a$ we have

$$du = (u_t + u_x a)dt = 0.$$

That is, u is constant along the curve given by $\frac{dx}{dt} = a$. This curve is called the *characteristic* of the partial differential equation. (If a is not a constant, then the characteristic is a curve and not a straight line.)

We next consider the characteristic equation

$$\frac{dx}{dt} = a,$$
$$x(0) = x_0,$$

whose solution is given by $x(t) = at + x_0$. Since $u(t, x)$ *must* be constant on this curve, we have

$$u(t, x) = u(0, x_0) = \Phi(x_0) = \Phi(x - at),$$

where the initial population density is given by $\Phi(x)$. The solution is determined by the initial condition which is moving to the right with velocity a (the "slope" of the characteristic curve) as t increases (see Figure 9.7).

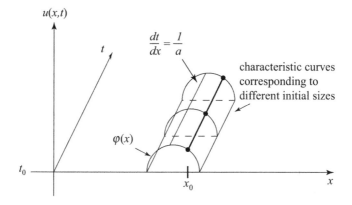

FIGURE 9.7:   Solution to equation (9.10) along the characteristic curve for $g(t, x) \equiv a$ and $\mu = 0$.

We now extend the above ideas to find the solution to the Sinko-Streifer model given by the following initial-boundary value problem:

$$\frac{\partial u}{\partial t} + \frac{\partial}{\partial x}(g(t,x)u(t,x)) = -\mu(t,x)u(t,x) \tag{9.12}$$

with boundary condition

$$g(t,x_0)u(t,x_0) = R(t) \tag{9.13}$$

and initial condition

$$u(0,x) = \Phi(x). \tag{9.14}$$

The total derivative of u along the characteristic curve $\frac{dx}{dt} = g$ is given by

$$\begin{aligned}
\frac{du}{dt} &= \frac{\partial u}{\partial x}\frac{dx}{dt} + \frac{\partial u}{\partial t}, \\
&= \frac{\partial u}{\partial x}g(t,x) + \frac{\partial u}{\partial t}.
\end{aligned}$$

Since $\frac{\partial u}{\partial t} + \frac{\partial}{\partial x}(g(t,x)u(t,x)) = -\mu(t,x)u(t,x)$, we obtain

$$\frac{\partial u}{\partial t} + g(t,x)\frac{\partial u}{\partial x} = -u(t,x)\frac{\partial}{\partial x}g(t,x) - \mu(t,x)u(t,x),$$

which implies

$$\frac{du}{dt} = -u(t,x)\frac{\partial}{\partial x}g(t,x) - \mu(t,x)u(t,x).$$

That is, along the characteristic curve $\frac{dx}{dt} = g(t,x)$, the solution of the Sinko-Streifer model satisfies the ordinary differential equation

$$\begin{aligned}
\frac{du}{dt} &= -ug_x - \mu u, \\
&= -(g_x + \mu)u,
\end{aligned}$$

which, after separation of variables, yields

$$u = v_0 e^{-\int (g_x + \mu)\,dt}. \tag{9.15}$$

Here, v_0 is a constant of integration yet to be determined.

We emphasize again that the solution $u = v_0 e^{-\int (g_x + \mu)dt}$ is valid only for (t,x) satisfying $\frac{dx}{dt} = g(t,x)$, that is, along the characteristic curve. Let $(t, X(t;\hat{t},\hat{x}))$ denote a characteristic curve passing through $(\hat{t},\hat{x})$ in the (t,x) plane as depicted in Figure 9.8, where X satisfies

$$\frac{d}{dt}X(t;\hat{t},\hat{x}) = g(t, X(t;\hat{t},\hat{x}))$$
$$X(\hat{t};\hat{t},\hat{x}) = \hat{x}.$$

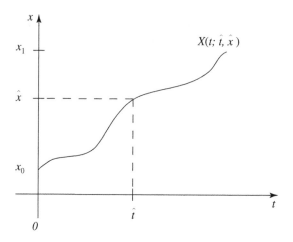

FIGURE 9.8: Characteristic curve.

The function $X(t; \hat{t}, \hat{x}) : t \to x$ maps t (time) to x (size). Since g is assumed to be positive, $\frac{dX}{dt} > 0$. Therefore, X is a strictly increasing function and hence has an inverse $T(x; \hat{t}, \hat{x}) : x \to t$. The characteristic curve passing through $(\hat{t}, \hat{x})$ is also given by $(T(x; \hat{t}, \hat{x}), x)$.

Now let $G(x) = T(x; 0, x_0)$ denote the curve passing through $(0, x_0)$. This curve divides the (t, x) plane into two parts as depicted in Figure 9.9. We therefore divide our considerations into two separate cases corresponding to the two regions in the (t, x) plane separated by the curve $(G(x), x) = (T(x; 0, x_0), x) = (t, X(t; 0, x_0))$.

1. For $t \le G(x)$ we obtain

$$u(t, x) = v_0 e^{-\int (g_x + \mu)\, dt}$$
$$= v_0 e^{-\int_0^t (g_x(\xi, x) + \mu(\xi, x))\, d\xi}. \qquad (9.16)$$

Evaluating equation (9.16) at $t = 0$ we find

$$\Phi(x) = u(0, x) = v_0 e^{-\int_0^0 (g_x + \mu)\, dt},$$

which then implies

$$v_0 = \Phi(x),$$

where $x = X(0; t, x)$. Substituting this back into equation (9.16) we have

$$u(t, x) = \Phi(x) e^{-\int_0^t [g_x(\xi, x) + \mu(\xi, x)]\, d\xi}.$$

This expression for $u(t, x)$ holds only for values of x and t that are on the characteristic curves. That is,

$$u(t, x) = \Phi(X(0; t, x)) e^{-\int_0^t [g_x(\xi, X(\xi; t, x)) + \mu(\xi, X(\xi; t, x))]\, d\xi}. \qquad (9.17)$$

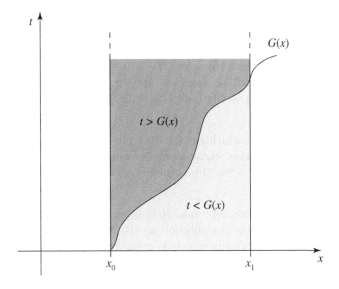

FIGURE 9.9: Regions in the (t, x) plane defining the solution.

2. Next, in the region where $t > G(x)$ we have

$$u(t, x) = v_0 e^{-\int_{T(x_0; t, x)}^{t} [g_x(\xi, x) + \mu(\xi, x)] d\xi}, \tag{9.18}$$

where $T(x_0; t, x)$ is the time corresponding to x_0. Hence at the point (t, x_0), $T(x_0; t, x) = t$. Therefore,

$$u(t, x_0) = v_0 e^{-\int_{T(x_0; t, x)}^{t} [g_x(\xi, x) + \mu(\xi, x)] d\xi}$$

$$= v_0 e^{-\int_{t}^{t} [g_x(\xi, x) + \mu(\xi, x)] d\xi}$$

$$= v_0.$$

We thus obtain from the boundary condition

$$v_0 = u(t, x_0) = \frac{R(t)}{g(t, x_0)}.$$

Substituting this into equation (9.18) we obtain

$$u(t, x_0) = \frac{R(t)}{g(t, x_0)} e^{-\int_{T(x_0; t, x)}^{t} [g_x(\xi, x) + \mu(\xi, x)] d\xi}.$$

Again, since this expression is only valid for (t, x) lying on the characteristic curves, we obtain

$$u(t, x) = \frac{R(T(x_0; t, x))}{g(T(x_0; t, x), x_0)} \times$$

$$e^{-\int_{T(x_0; t, x)}^{t} [g_x(\xi, X(\xi; t, x)) + \mu(\xi, X(\xi; t, x))] d\xi}. \tag{9.19}$$

Equations (9.17) and (9.19) define the complete analytical solution to the Sinko-Streifer model. In summary, this solution consists of two parts:

1. The first part in which the solution is given by (9.17) describes the part of the population whose members are survivors of the initial population density $u(0, x) = \Phi(x)$. This is called the *initial condition driven solution*.

2. The second part in which the solution is given by (9.19) describes the part of the population whose members were born after time $t = 0$ and enter the population via the boundary. This is called the *recruitment driven solution*.

We note that if the recruitment rate $R(t)$ is known, expressions (9.17) and (9.19) completely decouple the solution to the Sinko-Streifer equation. On the other hand, if the recruitment rate is given by (9.11), we have

$$R(T(x_0; t, x)) = \int_{x_0}^{x_1} k(T(x_0; t, x), \xi) u(T(x_0; t, x), \xi) d\xi \qquad (9.20)$$

and the solution does not decouple since values of u on the region $t < G(x)$ will be used to compute the recruitment rate. In the following two examples, we will consider two special cases where we show how one can apply the above derived formulas (9.17) and (9.19) to write explicitly the solutions to the Sinko-Streifer model (9.10).

Example: Constant Growth Rate and Mortality

In this example, we consider the simple case where

$$g = g_0 \text{ (constant)},$$
$$\mu = \mu_0 \text{ (constant)}.$$

Hence, the model is given by

$$\frac{\partial u}{\partial t} + g_0 \frac{\partial u}{\partial x} = -\mu_0 u$$

with boundary condition

$$u(t, x_0) g_0 = R(t)$$

and initial condition

$$u(0, x) = \Phi(x).$$

In this simple case, the characteristic curve passing through $(\hat{t}, \hat{x})$ is given by the initial-value problem

$$\frac{dx}{dt} = g_0$$
$$x(\hat{t}) = \hat{x},$$

which can be solved to obtain

$$x - \hat{x} = (t - \hat{t})g_0.$$

Thus, we obtain

$$X(t; \hat{t}, \hat{x}) = \hat{x} + g_0(t - \hat{t})$$
$$T(x; \hat{t}, \hat{x}) = \hat{t} + \frac{1}{g_0}(x - \hat{x}).$$

From these equations we can solve for the curve that passes through $(0, x_0)$ to be of the form:

$$G(x) = T(x; 0, x_0)$$
$$= \frac{1}{g_0}(x - x_0).$$

Therefore, the initial condition driven solution, which is the solution defined for $t \le G(x) = \frac{1}{g_0}(x - x_0)$, is given by

$$u(t, x) = \Phi(x - g_0 t)e^{-\int_0^t \mu_0 ds}$$
$$= \Phi(x - g_0 t)e^{-\mu_0 t}.$$

The recruitment driven solution, defined for $t > G(x) = \frac{1}{g_0}(x - x_0)$, has the form:

$$u(t, x) = \frac{R(T(x_0; t, x))}{g_0}e^{-\int_{T(x_0; t, x)}^t \mu_0 ds}$$
$$= \frac{R\left(t + \frac{1}{g_0}(x_0 - x)\right)}{g_0}e^{-\int_{t + \frac{1}{g_0}(x_0 - x)}^t \mu_0 ds}$$
$$= \frac{R\left(t + \frac{1}{g_0}(x_0 - x)\right)}{g_0}e^{\frac{\mu_0}{g_0}(x_0 - x)}.$$

In the next example, we will consider a more general case where the growth rate and the mortality are both size dependent.

Example: Size-Dependent Growth Rate and Mortality

In this example, we assume that

$$g = g(x) \text{ and } \mu = \mu(x).$$

Solving the characteristic equation

$$\frac{dx}{dt} = g(x)$$

by separation of variables and integration we obtain

$$\int_{x_0}^{x} \frac{dx}{g(x)} = \int_{0}^{t} dt = t.$$

Now defining the function

$$H(x) \equiv \int_{x_0}^{x} \frac{dx}{g(x)}$$

and observing that for $x_0 \leq \hat{x} \leq x$,

$$\int_{x_0}^{x} \frac{dx}{g(x)} = \int_{x_0}^{\hat{x}} \frac{dx}{g(x)} + \int_{\hat{x}}^{x} \frac{dx}{g(x)},$$

we find that

$$\int_{\hat{x}}^{x} \frac{dx}{g(x)} = \int_{x_0}^{x} \frac{dx}{g(x)} - \int_{x_0}^{\hat{x}} \frac{dx}{g(x)}.$$

Hence, we obtain

$$t = \hat{t} + \int_{\hat{x}}^{x} \frac{dx}{g(x)}$$

$$= \hat{t} + \int_{x_0}^{x} \frac{dx}{g(x)} - \int_{x_0}^{\hat{x}} \frac{dx}{g(x)}.$$

From the above equation, it follows that

$$T(x; \hat{t}, \hat{x}) = t$$
$$= \hat{t} + H(x) - H(\hat{x})$$

and

$$x = H^{-1}(t - \hat{t} + H(\hat{x})) = X(t; \hat{t}, \hat{x}).$$

In particular, recalling an earlier definition of $G(x) = T(x; 0, x_0)$, i.e., $(G(x), x)$ is the characteristic passing through $(0, x_0)$, we see that $T(x; 0, x_0) = 0 + H(x) - H(x_0) = H(x)$ so that in the case of time independent growth rate we have $G(x) = \int_{x_0}^{x} \frac{1}{g(\xi)} d\xi$. That is, $G(x) \equiv H(x)$. Thus, we have

$$T(x; \hat{t}, \hat{x}) = t$$
$$= \hat{t} + G(x) - G(\hat{x})$$

and

$$x = G^{-1}(t - \hat{t} + G(\hat{x}))$$
$$= X(t; \hat{t}, \hat{x}).$$

9.6 The Sinko-Streifer Model and Inverse Problems

So far, we have derived the analytical solution to the Sinko-Streifer model (9.12)-(9.14), given g, μ, R, and Φ. This is called the *forward problem*. Now we consider the *inverse problem*. That is, given $u(t,x)$ how do we find g, μ, R, and/or Φ?

One analytical method is due to Hackney and Webb [15], who developed a method for estimating growth and mortality rates from observed size distribution of larval fish, and demonstrated the method using data from larval crappie. We will now investigate the foundations of the Hackney-Webb method and identify as well any possible limitations.

Assume that we have observations $u(t_i, x_j)$ at times t_i for size x_j. In essence, when applying the Hackney-Webb method, one computes for each j the sums

$$n_j \equiv \sum_i u(t_i, x_j)$$

and

$$m_j \equiv \sum_i t_i u(t_i, x_j),$$

and then plots x_j versus the quotient $p_j = m_j/n_j$. From this, the growth g, which can depend on size, can be estimated. Similarly, a plot of n_j versus x_j yields an estimate of the mortality rate function μ. To explain this methodology, we introduce the functions $n(x)$, $m(x)$, $p(x)$ and an auxiliary function $h(x)$ given by

$$h(x) = \int_0^\infty (t - G(x))u(t,x)\,dt,$$

$$n(x) = \int_0^\infty u(t,x)\,dt,$$

$$m(x) = \int_0^\infty t u(t,x)\,dt,$$

$$p(x) = \frac{m(x)}{n(x)}.$$

We note that the quantities n_j, m_j and p_j may be viewed as discretizations for the integrals in n, m and p, respectively. To understand the foundations of the Hackney-Webb method, we need to develop some relationships between n, m, g, and μ.

We begin by noting that

$$h(x) = m(x) - G(x)n(x)$$

or

$$G(x) = \frac{m(x) - h(x)}{n(x)}$$

$$= p(x) - \frac{h(x)}{n(x)}.$$

If $\frac{h(x)}{n(x)}$ is constant, we obtain

$$G'(x) = \frac{1}{g(x)} = p'(x)$$

and hence a plot of x versus $p(x)$ will determine $g(x)$.

We next show that $\frac{h(x)}{n(x)}$ is constant in certain situations. Consider $n(x)$, where

$$n(x) = \int_0^\infty u(t,x)\,dt = \int_0^{G(x)} u(t,x)\,dt + \int_{G(x)}^\infty u(t,x)\,dt$$

$$= \int_0^{G(x)} u_{init}(t,x)\,dt + \int_{G(x)}^\infty u_{recr}(t,x)\,dt$$

$$= n_{init}(x) + n_{recr}(x),$$

where u_{init} and u_{recr} are the initial driven and recruitment driven solutions derived earlier, given by equations (9.17) and (9.19), respectively. That is, for $t \le G(x)$,

$$u_{init}(t,x) = \Phi(X(0;t,x))e^{-\int_0^t [g_x(X(\xi;t,x)) + \mu(X(\xi;t,x))]\,d\xi}$$

and for $t > G(x)$

$$u_{recr}(t,x) = \frac{R(T(x_0;t,x))}{g(x_0)}e^{-\int_{T(x_0;t,x)}^t [g_x(X(\xi;t,x)) + \mu(X(\xi;t,x))]\,d\xi},$$

where

$$T(x;\hat{t},\hat{x}) = \hat{t} + G(x) - G(\hat{x})$$

and

$$X(t;\hat{t},\hat{x}) = G^{-1}(G(\hat{x}) + t - \hat{t}). \tag{9.21}$$

Using the above equations for initial driven and recruitment driven solutions we now obtain

$$n_{init}(x) = \int_0^{G(x)} u_{init}(t,x)\,dt$$

$$= \int_0^{G(x)} \Phi(G^{-1}(G(x) - t))e^{-\int_0^t [g_x(X(\xi;t,x)) + \mu(X(\xi;t,x))]\,d\xi}\,dt.$$

Using the following substitutions

$$\eta = G^{-1}(G(x) - t)$$

and

$$s = X(\xi; t, x) = G^{-1}(G(x) + \xi - t)$$

we obtain

$$
\begin{aligned}
n_{init}(x) &= -\int_x^{x_0} \Phi(\eta) e^{-\int_\eta^x [g_x(s) + \mu(s)] G'(s)\, ds} G'(\eta)\, d\eta \\
&= \int_{x_0}^x \frac{\Phi(\eta)}{g(\eta)} e^{-\int_\eta^x \frac{1}{g(x)} [g_x(s) + \mu(s)]\, ds}\, d\eta,
\end{aligned}
$$

where we use the identity $G(x) = \int_{x_0}^x \frac{dx}{g(x)}$ and equation (9.21). Similarly, for the recruitment driven term

$$
\begin{aligned}
n_{recr}(x) &= \int_{G(x)}^\infty u_{recr}(t, x)\, dt \\
&= \int_{G(x)}^\infty \frac{R(T(x_0; t, x))}{g(x_0)} e^{-\int_{T(x_0;t,x)}^t [g_x(X(\xi;t,x)) + \mu(X(\xi;t,x))]\, d\xi}\, dt,
\end{aligned}
$$

if we let $\sigma = T(x_0; t, x) = t + G(x_0) - G(x)$, we obtain

$$n_{recr}(x) = \int_0^\infty \frac{1}{g(x_0)} R(\sigma) e^{-\int_{x_0}^x \frac{1}{g(s)} [g_x(s) + \mu(s)]\, ds}\, d\sigma.$$

Now letting

$$D(x) = e^{-\int_{x_0}^x \frac{1}{g(s)} [g_x(s) + \mu(s)] ds}$$

we find

$$
\begin{aligned}
n_{recr}(x) &= D(x) \int_0^\infty \frac{R(\sigma)}{g(x_0)}\, d\sigma \\
&= c_1 D(x),
\end{aligned}
$$

where c_1 is a constant. Similarly, it can be shown that

$$n_{init} = D(x) \int_{x_0}^x \frac{\Phi(\eta)}{g(\eta)} e^{+\int_{x_0}^\eta \frac{1}{g(s)} [g_x(s) + \mu(s)]\, ds}\, d\eta,$$

$$
\begin{aligned}
h_{init}(x) &= -\int_{x_0}^x G(\eta) \frac{\Phi(\eta)}{g(\eta)} e^{-\int_\eta^x \frac{1}{g(s)} [g_x(s) + \mu(s)]\, ds}\, d\eta \\
&= -D(x) \int_{x_0}^x G(\eta) \frac{\Phi(\eta)}{g(\eta)} e^{+\int_{x_0}^\eta \frac{1}{g(s)} [g_x(s) + \mu(s)]\, ds}\, d\eta,
\end{aligned}
$$

$$h_{recr}(x) = D(x) \int_0^\infty \sigma \frac{R(\sigma)}{g(x_0)}\, d\sigma.$$

Combining the above results, we obtain

$$
\begin{aligned}
n(x) &= n_{init}(x) + n_{recr}(x) \\
&= D(x) \int_{x_0}^{x} \frac{\Phi(\eta)}{g(\eta)} e^{\int_{x_0}^{\eta} \frac{1}{g}[g_x + \mu]\, ds}\, d\eta + D(x)c_1 \\
&= D(x) \left[c_1 + \int_{x_0}^{x} \frac{\Phi(\eta)}{g(\eta)} e^{\int_{x_0}^{\eta} \frac{1}{g}[g_x + \mu]\, ds}\, d\eta \right]
\end{aligned}
$$

while

$$
\begin{aligned}
h(x) &= h_{init}(x) + h_{recr}(x) \\
&= D(x) \left[c_2 - \int_{x_0}^{x} G(\eta) \frac{\Phi(\eta)}{g(\eta)} e^{\int_{x_0}^{\eta} \frac{1}{g}[g_x + \mu]\, ds}\, d\eta \right].
\end{aligned}
$$

Hence, if Φ vanishes outside $[x_0, \bar{x}]$, then for $x > \bar{x}$ we have

$$
\begin{aligned}
\eta(x) &= D(x)c_1, \\
h(x) &= D(x)c_2.
\end{aligned}
$$

So $\frac{h(x)}{\eta(x)} = \frac{c_2}{c_1}$, which is a constant for $x \geq \bar{x}$ if Φ vanishes outside $[x_0, \bar{x}]$. Hence, the Hackney-Webb method can be expected to give estimates for zero initial conditions in case of time independent growth and mortality!

A more complete comparison of the method of Hackney and Webb to the inverse least squares method introduced in the project below can be found in [8].

9.7 Size Structure and Mosquitofish Populations

We return now to the mosquitofish populations that we introduced as motivation at the beginning of this chapter. We have discussed in previous sections the Sinko-Streifer model and methods for its solution in both forward problem and inverse problem settings. While the Sinko-Streifer model is widely (and successfully) used in the literature (see [1, 2, 17]) on biological populations, it has some rather serious shortcomings. These are readily seen in considering the mosquitofish data [12] depicted in Figure 9.10. In this data, we see that a pulse of population (23 July) exhibits in time both dispersion (6 August) and bifurcation (25 August). That is, a unimodal density disperses and becomes bimodal. Recalling the solution of the Sinko-Streifer equation (in particular, the initial condition driven solution (9.17)), we see that a pulse propagates without dispersion or bifurcation. The initial data Φ propagates along characteristics emanating from its region of non-zero support with amplitude increasing or decreasing in time depending on the values of $\frac{\partial g}{\partial x}$ and μ.

In fact, one can argue from the Sinko-Streifer equation itself and the method of characteristics that dispersion cannot occur unless $\frac{\partial g}{\partial x} > 0$; this is a condition for spreading of the characteristic curves defined by $\frac{dx}{dt} = g$. Such an assumption is inherently unreasonable in many biological applications: in our example it is equivalent to the assumption that individual growth rates increase as one's size increases! Indeed, it is counter-intuitive that the larger one is, the faster one grows.

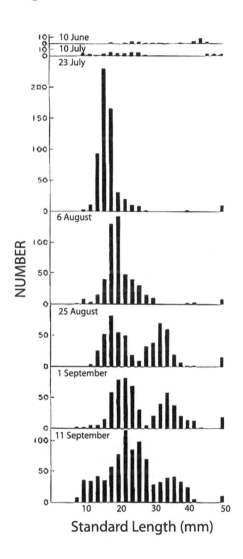

FIGURE 9.10: Mosquitofish data.

Thus, while there is no hope that the Sinko-Streifer model as developed in the previous sections can describe the mosquitofish data, one might be reluctant to abandon such a popular as well as reasonable growth model. One might instead turn to a more careful analysis of the assumptions underlying the Sinko-Streifer equation. Further investigation of the mosquitofish populations and their biological properties leads to the additional information that males and females reach different maximum sizes (30mm and 60mm, respectively). Hence, we conclude that males and females in the size range 28-30mm must grow at different rates. This immediately violates one of the underlying assumptions of the Sinko-Streifer model that individuals of the same size grow at the same rate, i.e., the assumption $\frac{dx}{dt} = g(t, x)$ cannot be reasonable for mosquitofish (and possibly other) data. We next describe an idea first introduced in [9] and later theoretically developed in [4, 5, 6, 10] and more recently in modeling of early growth of shrimp [3, 7].

To generalize the Sinko-Streifer equation, we allow individuals of the same size to possess different individual growth rates. This can be accomplished by assuming the existence of "intrinsic" parameters γ (which in general we cannot observe and hence cannot use to physically distinguish individuals in the data) on which the growth rates depend. Thus we assume

$$\frac{dx}{dt} = g(t, x; \gamma) \tag{9.22}$$

as a parameter-dependent individual growth rate. The parameter values may range over a set of admissible parameters Γ, and the total population is composed of subgroups, grouped together in population substructures characterized by common γ values. For example, if $\Gamma = \{\gamma_1, \gamma_2\}$ (think males and females with each gender possessing a different γ value), and p_i is the proportion of individuals with intrinsic parameters γ_i, then the total population density $v(t, x)$ would be given by

$$v(t, x) = p_1 u(t, x; \gamma_1) + p_2 u(t, x; \gamma_2), \tag{9.23}$$

where $u(t, x; \gamma_i)$ is the solution to the Sinko-Streifer equation using (9.22) with $\gamma = \gamma_i$.

Of course, as soon as one admits the generalization, it is quite reasonable to assume multiple subclasses corresponding to a finite (or even infinite) family of γ values (again, think here that not all males of the same size have the same γ values), leading to a *distribution of growth rates* within the population. For $\Gamma = \{\gamma_1, \cdots, \gamma_M\}$ with corresponding proportions (probabilities) p_i, with $\sum p_i = 1$, the expression (9.23) for total population density generalizes to

$$v(t, x) = \sum_{i=1}^{M} p_i u(t, x; \gamma_i). \tag{9.24}$$

In the case of an (infinite) continuum Γ of intrinsic parameter values, the above ideas generalize to a probability measure or distribution P characterizing a distribution of the γ's in Γ. Equation (9.24) becomes

$$v(t, x) = \int_\Gamma u(t, x; \gamma) dP(\gamma). \tag{9.25}$$

If the distribution P is (absolutely) continuous, i.e., possesses a corresponding density $p = \frac{dP}{d\gamma}$, then one has

$$v(t, x) = \int_\Gamma u(t, x; \gamma) p(\gamma) d\gamma.$$

In [11] it is shown that these generalizations of Sinko-Streifer equation to include distributed growth rates do indeed allow the required dispersion and bifurcation so as to describe well the mosquitofish data depicted in Figure 9.10. In the exercise on distributed growth rates below the requested simulations allow the reader to demonstrate these features of the generalized Sinko-Streifer model. More recently [3, 7], the Sinko-Streifer system with growth rate distributions has been successfully used to model the variability in the early growth of shrimp. A mathematical and stochastic theoretical foundation as well as computational ideas for this formulation can be found in [3, 4, 5, 6, 10]

Project: Size-Structured Population Model Inverse Problem

Consider the following Sinko-Streifer model for a larval fish population:

$$\frac{\partial u}{\partial t} + \frac{\partial}{\partial x}(gu) = -\mu u, \qquad t \in (0, t_f], \qquad x \in (x_0, x_1)$$

with boundary condition

$$g(t, x) u(t, x)|_{x=x_0} = R(t)$$

and initial condition

$$u(0, x) = \Phi(x).$$

1. In the case of constant growth and death rates ($g(t, x) = g_0$ and $\mu(t, x) = \mu_0$), derive the exact solution $u(t, x)$ by the method of characteristics.

2. Define the recruitment function $R(t)$ to be as follows:

$$R(t) = \begin{cases} 3/4((\alpha t - 1) - (\alpha t - 1)^3/3 + 2/3), & t \in [0, 2/\alpha], \\ 1, & t \in [2/\alpha, 2/\alpha + \beta], \\ -3/4((\alpha s - 1) - (\alpha s - 1)^3/3 - 2/3), & s = t - 2/\alpha - \beta, \\ & t \in [2/\alpha + \beta, 4/\alpha + \beta], \\ 0, & \text{otherwise}, \end{cases}$$

where $\alpha = 15$ and $\beta = 1/\alpha$.

(i) Plot the recruitment function $R(t)$ as a function of t for $t \in [0,5]$.

(ii) Assume that the sampling of the population starts prior to the beginning of the reproductive season. That is the initial population distribution is zero ($\Phi(x) = 0$). (This will also allow us to utilize the method of Hackney and Webb to determine the growth rate g and the death rate μ.) Furthermore, assume that $x_0 = 0$, $x_1 = 0.5$, growth rate $g_0 = 0.185$ and death rate $\mu_0 = 1.9$. Plot the exact solution $u(t, x)$ of the Sinko-Streifer model as a function of $t \in [0,5]$ and x (three-dimensional plot). Describe the dynamics of $u(t, x)$.

3. We now create *"simulated"* data to be used for estimating growth and mortality rates. For this, we assume that population is sampled at equally spaced time and size intervals. We will subdivide the size interval $[0, 0.5]$ and time interval $[0, 5]$ into N_x and N_t equal subintervals of length $h_x = 0.5/N_x$ and $h_t = 5/N_t$ respectively. Let $u_s(t_i, x_j)$ denote the number of larval fish sampled at size $x_j = j \cdot h_x$, $j = 1, \ldots, N_x$ and at time $t_i = i \cdot h_t$, $i = 1, \ldots, N_t$. For the following problem take $N_x = N_t = 20$.

(i) Use the Hackney and Webb method to estimate the growth and death rates g_0, μ_0 respectively.

(ii) Now repeat the above question but using the inverse least squares technique. For this question we will use the following least squares criteria: Minimize

$$J(g, \mu) = \sum_{i=1}^{N_t} \sum_{j=1}^{N_x} |u_c(t_i, x_j; g, \mu) - u_s(t_i, x_j)|^2,$$

where $u_c(t_i, x_j; g, \mu)$ is the characteristic solution evaluated at size x_j and time t_i for given values of g and μ. To compute the values of g_0^* and μ_0^* which minimize the above least squares criteria, use the MATLAB routine `fminu` or the Nelder-Mead algorithm `fminsearch`.

4. In practice, the data collected is corrupted by noise (for example, errors in collecting data, instrumental errors, etc.). In the following exercise, we will test the sensitivity of the method of Hackney and Webb and the inverse least squares method to errors in sampling the data. For this, we will add to each simulated data an error term as follows:

$$\hat{u}_s(t_i, x_j) = u_s(t_i, x_j) + nl \cdot rand_{ij},$$

where $rand_{ij}$ are the normally distributed random numbers with zero mean and variance 1.0. Use the MATLAB routine `randn(m,n)` to generate an m-by-n matrix with random entries. Finally, nl is a noise level constant.

(i) For $nl = 0.01, 0.02, 0.05, 0.1, 0.2$, estimate the growth and death rates using both the Hackney and Webb method and the inverse least squares method. Create a table listing the estimate values of the growth and death rates and the values of the cost functionals for each value of nl. Describe the sensitivity of each method with respect to the noise level nl.

(ii) Plot on the same graph the solution $u(t, x)$ computed using the exact values of g_0 and μ_0 and the estimated values of g_0 and μ_0 for $nl = 0.01$.

Project: Distributed Growth Rate Population Model

Consider the following Sinko-Streifer model for a size-structured mosquitofish population:

$$\frac{\partial u}{\partial t} + \frac{\partial}{\partial x}(gu) = -\mu u, \qquad t \in [0, t_f], \qquad x \in [x_0, x_1]$$

with boundary condition

$$g(t, x)u(t, x)|_{x=x_0} = R(t)$$

and initial condition

$$u(0, x) = \Phi(x).$$

Use the following initial and terminal size classes, initial density and recruitment rate throughout:

- $x_0 = 0, \quad x_1 = 1,$

- $\Phi(x) = \begin{cases} \sin^2 10\pi x, & 0 \le x \le 0.1, \\ 0, & 0.1 < x \le 1, \end{cases}$

- $R(t) = 0.$

1. In the case of size-dependent growth rate and constant death rate (i.e., $g(t, x) = g(x)$ and $\mu(t, x) = \mu_0$), derive the exact solution $u(t, x)$ using the method of characteristics. Consider the case when $g(x) = b(1 - x)$ where b is the intrinsic growth rate of the mosquitofish. Furthermore, assume $b = 4.5$ while considering the three different cases: $\mu_0 = 0$, $\mu_0 = 4.5$, and $\mu_0 = 7.5$.

   (a) Plot (3D plot) the exact solution $u(t, x)$ as a function of $t \in [0, 0.5]$, and $x \in [0, 1]$.

(b) Plot (2D plot) the exact solution $u(t_i, x)$ versus x for several different moments in time t_i.

Do this for all three values of μ_0.

2. As discussed previously, assuming that same sized individuals grow at the same rate is biologically unreasonable. Indeed, one would suspect that individuals in a population have intrinsic parameters that affect their growth rates. Included in these parameters is the intrinsic growth rate, b, in the above example, of the mosquitofish. Consider populations with a Gaussian distribution on b with a mean of 4.5 and variance of 0.0816, i.e., b is $\mathcal{N}(\bar{b}, \sigma^2)$ with $\bar{b} = 4.5$ and $\sigma^2 = 0.0816$, and again carry out the computations below for each of the three values of μ_0 given above.

(a) Plot (3D plot) the exact solution $u(t, x)$ as a function of $t \in [0, 0.5]$, and $x \in [0, 1]$.

(b) Plot (2D plots) the exact solution $u(t_i, x)$ versus $x \in [0, 1]$ for several different moments in time, t_i.

(c) Do these solutions differ from those in part 1.? If so, how?

3. Also as discussed previously, the mosquitofish population data appears to consist of at least two subclasses with differing growth characteristics. Thus, in order to attempt to simulate a similar population, consider the case where $g_1(x) = b_1(1 - x)$ and $g_2(x) = b_2(1 - x)$ for the two distinct subclasses.

(a) Assume a Bi-Gaussian distribution on b with an overall mean of 4.5, with subpopulation means $\bar{b}_1 = 3.3$ and $\bar{b}_2 = 5.7$ as well as variance $\sigma^2 = 0.492$ each. As above, carry out the below for each of the three values of μ_0 given in part 1.

  i. Plot the exact solution $u(t, x)$ as a function of $t \in [0, 0.5]$, and $x \in [0, 1]$.

  ii. Plot the exact solution $u(t_i, x)$ versus x for several different moments in time, t_i.

  iii. Do these solutions differ from those in questions 1 and 2? If so, how?

(b) In the previous questions, we assumed a constant maximum size with no dependence on the intrinsic growth rates. However, it is more reasonable to assume that the maximum size is dependent on the intrinsic growth rates of the mosquitofish. Letting S denote the maximum size, we let $S(b) = 0.6 + \frac{4}{11}(b - 3.3)$. Then consider a Bi-Gaussian distribution on b with overall mean of $\bar{b} = 4.4$ with subpopulation means $\bar{b}_1 = 3.8$ and $\bar{b}_2 = 5.0$ as well as variance $\sigma^2 = 0.123$ each. Again do this for all three values of μ_0.

i. Plot the exact solution $u(t, x)$ as a function of $t \in [0, 0.5]$ and x.

ii. Plot the exact solution $u(t_i, x)$ versus x for several different moments in time, t_i.

iii. Discuss the results (e.g., compare with the results of 1, 2, 3a).

References

[1] H.T. Banks, J.E. Banks, L.K. Dick and J.D. Stark, Estimation of dynamic rate parameters in insect populations undergoing sublethal exposure to pesticides, CRSC-TR05-22, NCSU, May, 2005; *Bull. Math. Biol.*, **69** (2007), pp. 2139–2180.

[2] H.T. Banks, J.E. Banks, L.K. Dick and J.D. Stark, Time-varying vital rates in ecotoxicology: selective pesticides and aphid population dynamics, *Ecological Modelling*, **210** , 2008, pp. 155–160.

[3] H.T. Banks, V.A. Bokil, S. Hu, F.C.T. Allnutt, R. Bullis, A. K. Dhar and C. L. Browdy, Shrimp biomass and viral infection for production of biological countermeasures, CRSC-TR05-45, NCSU, December, 2005; *Mathematical Biosciences and Engineering*, **3**, 2006, pp. 635–660.

[4] H.T. Banks, D.M. Bortz, G.A. Pinter and L.K. Potter, Modeling and imaging techniques with potential for application in bioterrorism, CRSC-TR03-02, NCSU, January, 2003; Chapter 6 in Bioterrorism: Mathematical Modeling Applications in Homeland Security, (H.T. Banks and C. Castillo-Chavez, eds.), Frontiers in Applied Math, **FR28**, SIAM, Philadelphia, PA, 2003, pp. 129–154.

[5] H.T. Banks and J.L. Davis, A comparison of approximation methods for the estimation of probability distributions on parameters, CRSC-TR05-38, NCSU, October, 2005; *Applied Numerical Mathematics*, **57**, 2007, pp. 753-777.

[6] H.T. Banks and J.L. Davis, Quantifying uncertainty in the estimation of probability distributions with confidence bands, CRSC-TR07-21, NCSU, December, 2007; *Math. Biosci. Engr.*, **5**, 2008, pp. 647–667.

[7] H.T. Banks, J.L. Davis, S.L. Ernstberger, S. Hu, E. Artimovich, A.K. Dhar and C.L. Browdy, Comparison of probabilistic and stochastic formulations in modeling growth uncertainty and variability, CRSC-TR08-03, NCSU, February, 2008; *J. Biological Dynamics*, to appear.

[8] H.T. Banks, L.W. Botsford, F. Kappel and C. Wang, Estimation of growth and survival in size-structured cohort data: An application to larval striped bass (Morone Saxatilis), *J. Math. Biol.*, **30**, 1991, pp. 125–150.

[9] H.T. Banks, L.W. Botsford, F. Kappel and C. Wang, Modeling and estimation in size-structured population models, in *Proceeding 2nd Course on Mathematical Ecology*, Trieste, 1986, World Press, Singapore, 1988, pp. 521–541.

[10] H.T. Banks and B.G. Fitzpatrick, Estimation of growth rate distributions in size-structured population models, *Quart. Appl. Math.*, **49**, 1991, pp. 215–235.

[11] H.T. Banks, B.G. Fitzpatrick, L.K. Potter and Y. Zhang, Estimation of probability distributions for individual parameters using aggregate population observations, in *Stochastic Analysis, Control, Optimization and Applications* (W.Mceneaney, G. Yin, Q. Zhang, eds.), Birkhäuser, 1998, pp. 353–371.

[12] L.W. Botsford, B. Vandracek, T. Wainwright, A. Linden, R. Kope, D. Reed and J.J. Cech, Population development of the mosquitofish, Gambusia Affinis, in rice fields, *Environ. Biol. Fishes*, **20**, 1987, pp. 143–154.

[13] M. Braun, *Differential Equations and Their Applications: An Introduction to Applied Mathematics*, Springer, Berlin, 4th ed., 1992.

[14] R. Courant and D. Hilbert, *Methods of Mathematical Physics, vol. II*, Wiley, New York, 1962.

[15] P.A. Hackney and J.C. Webb, A method for determining growth and mortality rates of ichthyoplankton, in *Proc. Fourth National Workshop on Entrainment and Impringement* (L.D. Jenson, ed.), Ecological Analysts Inc., Melville, New York, 1978, pp. 115–124.

[16] W. Kaplan, *Advanced Calculus*, Addison-Wesley Publishing Co., Inc., New York, 1991.

[17] M. Kot, *Elements of Mathematical Ecology*, Cambridge University Press, Cambridge, 2001.

[18] J.A.J. Metz and O. Diekmann (eds.), *The Dynamics of Physiologically Structured Populations*, Lecture Notes in Biomathematics, **68**, Springer, 1986.

[19] R. Pearl and L.J. Reed, On the rate of growth of the population of the United States since 1790 and its mathematical representation, *Proceedings of the National Academy of Sciences of the United States of America*, **6**, pp. 275–288.

[20] J.W. Sinko and W. Streifer, A new model for age-size structure of a population, *Ecology*, **48**, 1967, pp. 910–918.

[21] P.F. Verhulst, Recherches mathématiques sur la loi d'accroissement de la population, *Noveaux Méoires de l'Académie Royale des Sciences et Belles Lettres de Bruxelles*, **18**, pp. 3–38.

[22] V. Volterra, Fluctuation in the abundance of a species considered mathematically, *Nature*, **118**, pp. 558–560.

Appendix A

An Introduction to Fourier Techniques

In this appendix we review the basic tools and techniques from linear system theory used in the analysis of periodic and non-periodic waveforms. In particular, the development of Fourier methods has had a major impact on the analysis of linear systems. It allows the analysis of complex waveforms by considering sinusoidal components (see, for instance, [3, 4]).

A.1 Fourier Series

Functions that are periodic with finite energy within each period can be represented by a *Fourier series*. That is, any real, periodic function $x(t)$ can be represented as an infinite sum of increasing harmonic sine and cosine components as

$$x(t) = \frac{1}{2}a_0 + \sum_{n=1}^{\infty}(a_n \cos(nt) + b_n \sin(nt)). \qquad (A.1)$$

Here, the terms $\cos(t)$ and $\sin(t)$ are called the *fundamental terms*. The higher component terms $\cos(nt)$ and $\sin(nt)$, for integer $n > 1$, are called *harmonic terms*.

The calculation of the coefficients a_n and b_n are facilitated by the following properties of the sine, cosine, their products and cross-products.

For any integers m and n,

$$\int_{-\pi}^{\pi} \sin mt \, dt = 0,$$

$$\int_{-\pi}^{\pi} \cos mt \, dt = 0,$$

$$\int_{-\pi}^{\pi} \sin mt \cos nt \, dt = 0. \qquad (A.2)$$

For integers $m \neq n$,

$$\int_{-\pi}^{\pi} \sin mt \sin nt \, dt = 0,$$

$$\int_{-\pi}^{\pi} \cos mt \cos nt \, dt = 0. \tag{A.3}$$

For integers $m = n$,

$$\int_{-\pi}^{\pi} (\sin mt)^2 \, dt = \pi,$$

$$\int_{-\pi}^{\pi} (\cos mt)^2 \, dt = \pi. \tag{A.4}$$

Using the above formulas (A.2)-(A.4), we obtain the coefficient a_0 by integrating both sides of equation (A.1) from $-\pi$ to π to yield

$$a_0 = \frac{1}{\pi} \int_{-\pi}^{\pi} x(t) \, dt. \tag{A.5}$$

Similarly, by multiplying both sides of equation (A.1) by $\sin nt$ and $\cos nt$ and integrating, we obtain

$$a_n = \frac{1}{\pi} \int_{-\pi}^{\pi} x(t) \cos(nt) \, dt, \tag{A.6}$$

$$b_n = \frac{1}{\pi} \int_{-\pi}^{\pi} x(t) \sin(nt) \, dt, \tag{A.7}$$

respectively.

The type of Fourier series expressed by equation (A.1) is known as a *trigonometric Fourier series* and can be applied only to real, periodic functions. Another form of Fourier series, known as the *exponential Fourier series*, can be applied to both real-valued and complex-valued functions $x(t)$ as long as they are periodic. This form of Fourier series makes use of the following identities

$$\cos t = \frac{e^{jt} + e^{-jt}}{2}, \tag{A.8}$$

$$\sin t = \frac{e^{jt} - e^{-jt}}{2j}. \tag{A.9}$$

In addition, for a periodic function with period T_0, the frequency component $f_0 = \frac{1}{T_0}$ is called the *fundamental frequency*. The higher frequency component $f_n = nf_0$, for $n > 1$ and integer, is called the *nth harmonic*. Another frequency component denoted by $\omega_0 = 2\pi/T_0$ is called the fundamental *radian* frequency. Both terms f and ω are used to denote frequency. When f is used, the unit of frequency is in hertz (Hz); when ω is used, frequency in radians/second is intended.

By replacing the nt term in equation (A.1) with $2\pi n f_0 t = 2\pi n t / T_0$ and using the identities (A.9) we can express $x(t)$ in exponential form as follows:

$$x(t) = \frac{1}{2}a_0 + \frac{1}{2}\sum_{n=1}^{\infty}\left[(a_n - jb_n)e^{j2\pi n f_0 t} + (a_n + jb_n)e^{-j2\pi n f_0 t}\right]. \qquad (A.10)$$

The above expression can be rewritten in the following form

$$x(t) = \sum_{n=-\infty}^{\infty} c_n e^{j2\pi n f_0 t}, \qquad (A.11)$$

where the complex coefficients c_n are defined by

$$c_n = \begin{cases} \frac{1}{2}a_0 & n = 0 \\ \frac{1}{2}(a_n - jb_n) & n > 0 \\ \frac{1}{2}(a_n + jb_n) & n < 0 \end{cases}$$

and can be computed from the following equation

$$c_n = \frac{1}{T_0}\int_{-T_0/2}^{T_0/2} x(t)e^{-j2\pi n f_0 t}\, dt.$$

The complex Fourier coefficient c_n can be expressed by the following expression

$$c_n = |c_n|e^{j\theta_n}, \qquad (A.12)$$
$$c_{-n} = |c_n|e^{-j\theta_n}, \qquad (A.13)$$

where

$$|c_n| = \frac{1}{2}\sqrt{a_n^2 + b_n^2}, \qquad (A.14)$$

$$\theta_n = \tan^{-1}\left(-\frac{b_n}{a_n}\right), \qquad (A.15)$$

and $b_0 = 0$ and $c_0 = \frac{a_0}{2}$. Plots of $|c_n|$ and θ_n versus n or nf_0 are called the *discrete spectra* of $x(t)$. The plot of $|c_n|$ is usually called the *magnitude spectrum*, and the plot of θ_n is referred to as *phase spectrum*. Furthermore, if $x(t)$ is a real-valued periodic time function, we have

$$c_{-n} = c_n^*,$$

which implies that the magnitude spectrum is an even function of frequency (since $|c_n| = |c_{-n}|$). Similarly, from equation (A.15), the phase spectrum is an odd function of frequency.

## A.2	Fourier Transforms

The Fourier series is very useful in characterizing arbitrary periodic functions or waveforms. For nonperiodic signals, however, a frequency-domain approach based on the Fourier transforms provides a more convenient representation.

Let $x(t)$ denote an arbitrarily absolutely integrable function. That is,

$$\int_{-\infty}^{\infty} |x(t)|\, dt < \infty.$$

The Fourier transform of $x(t)$, denoted by $X(f)$ or, equivalently, $\mathcal{F}[x(t)]$, is defined by

$$\mathcal{F}[x(t)] = X(f) = \int_{-\infty}^{\infty} x(t)e^{-j2\pi ft}\, dt. \tag{A.16}$$

The inverse Fourier transform of $X(f)$ is $x(t)$ and is given by

$$\mathcal{F}^{-1}[X(f)] = x(t) = \int_{-\infty}^{\infty} X(f)e^{j2\pi ft}\, df.$$

It is noted that the Fourier transform is a function of the function values on $(-\infty, \infty)$ and, consequently, the initial conditions are not treated. This differs from the Laplace transform. Since the Laplace transform of $x(t)$ is defined by

$$\mathcal{L}[x(t)] = \int_{0}^{\infty} e^{-st}x(t)\, dt,$$

it involves the initial condition effects in a nontrivial way.

We now summarize the most important properties of the Fourier transform.

1. *Differentiation*: Differentiation in the time domain corresponds to multiplication by $j2\pi f$ in the frequency domain. That is,

$$\mathcal{F}[x'(t)] = j2\pi f X(f),$$
$$\mathcal{F}[x^{(n)}(t)] = (j2\pi f)^n X(f).$$

2. *Modulation*: Multiplication by $e^{j2\pi f_0 t}$ in the time domain is equivalent to a frequency shift in the frequency domain:

$$\mathcal{F}[e^{j2\pi f_0 t}x(t)] = X(f - f_0).$$

3. *Time shift*: A time shift in the time domain results in a phase shift in the frequency domain:

$$\mathcal{F}[x(t - t_0)] = e^{-j2\pi f t_0}X(f).$$

4. *Duality*:
$$\mathcal{F}[X(t)] = x(-f).$$

5. *Hermitian Symmetry*: If $x(t)$ is a real-valued function,
$$X(-f) = X^*(f).$$

6. *Convolution*: Convolution in the time domain corresponds to a multiplication in the frequency domain, and vice versa. That is,
$$\mathcal{F}[x(t) * y(t)] = X(f)Y(f)$$
$$\mathcal{F}[x(t)y(t)] = X(f) * Y(f),$$

where $Y(f) = \mathcal{F}[y(t)]$ and $x(t) * y(t) = \int_{-\infty}^{\infty} x(t - \tau)y(\tau)\, d\tau$.

7. *Parseval's Indentity*:
$$\int_{-\infty}^{\infty} x(t)y^*(t)\, dt = \int_{-\infty}^{\infty} X(f)Y^*(f)\, df.$$

In the case that $y(t) = x(t)$, we obtain *Rayleigh's relation*
$$\int_{-\infty}^{\infty} |x(t)|^2\, dt = \int_{-\infty}^{\infty} |X(f)|^2\, df.$$

As already mentioned above, the Fourier transform is the extension of the Fourier series concept to non-periodic functions. However, for a periodic function $x(t)$, with period T_0, the exponential Fourier series expansion is given by

$$x(t) = \sum_{n=-\infty}^{\infty} c_n e^{j2\pi nt/T_0}. \tag{A.17}$$

Now, applying the Fourier transform to both sides of the above equation, we obtain

$$X(f) = \mathcal{F}\left[\sum_{n=-\infty}^{\infty} c_n e^{j2\pi nt/T_0}\right]$$
$$= \sum_{n=-\infty}^{\infty} c_n \mathcal{F}[e^{j2\pi nt/T_0}]$$
$$= \sum_{n=-\infty}^{\infty} c_n \delta\left(f - \frac{n}{T_0}\right).$$

In other words, the Fourier transform of a periodic function consists of impulses at multiples of the fundamental frequency (harmonics) of the original signal. Because of this property it is used widely in signal processing methodology.

Appendix B

Review of Vector Calculus

This appendix is devoted to a brief presentation of the essential aspects of vector calculus. In particular, we will review results that relate line integrals to double integrals and triple integrals to surface integrals. The corresponding formulas of Stokes and Gauss are of great importance in many physical and engineering problems and are presented here without proofs. For an in-depth treatment of this subject, we refer the interested reader to many excellent textbooks in the literature (see, e.g., [1, 2, 4]).

We begin by recalling some properties of the algebraic operations and combinations of vector fields. These properties occur in many theoretical and application considerations.

1. *Curl of a gradient*: An important corresponding formula is

$$\text{curl grad } f = \vec{0},$$

   or $\nabla \times (\nabla f) = 0$. Conversely, if curl $\vec{u} = 0$, then $\vec{u} = \text{grad } f$ for some scalar function f. A vector field $\vec{u}$ such that curl $\vec{u} = \vec{0}$ is said to be *irrotational*.

2. *Divergence of a curl*: One also has the rule

$$\text{div curl } \vec{u} = 0.$$

   Similarly, there is also an important converse; that is, if div $\vec{u} = 0$, then $\vec{u} = \text{curl } \vec{v}$ for some vector field $\vec{v}$. A vector field $\vec{u}$ such that div $\vec{u} = 0$ is often termed *solenoidal*.

3. *Divergence of a gradient*: Here one has the expression

$$\text{div grad } f = \frac{\partial^2 f}{\partial x^2} + \frac{\partial^2 f}{\partial y^2} + \frac{\partial^2 f}{\partial z^2}.$$

   The expression on the right side of the above equation is known as the *Laplacian* of f and is denoted by $\triangle f$ or $\nabla^2 f$. A function f with continuous second partial derivatives such that $\nabla^2 f = 0$ in a domain is called *harmonic* in that domain and the equation $\nabla^2 f = 0$ is called *Laplace's equation*.

4. *Divergence of a vector product:*

$$\text{div } (\vec{u} \times \vec{v}) = \vec{v} \cdot \text{curl } \vec{u} - \vec{u} \cdot \text{curl } \vec{v}.$$

5. *Curl of a vector product:*

$$\text{curl } (\vec{u} \times \vec{v}) = (\text{div } \vec{v})\vec{u} - (\text{div } \vec{u})\vec{v} + (\vec{v} \cdot \nabla)\vec{u} - (\vec{u} \cdot \nabla)\vec{v}.$$

6. *Curl of a curl:* Another important property is the relation

$$\text{curl curl } \vec{u} = \text{grad div } \vec{u} - \nabla^2 \vec{u},$$

where the Laplacian of a vector field $\vec{u} = (u_x, u_y, u_z)$ is defined to be

$$\nabla^2 \vec{u} = \nabla^2 u_x \vec{i} + \nabla^2 u_y \vec{j} + \nabla^2 u_z \vec{k}.$$

We now review some important topics of vector integral calculus. First, some definitions are in order. A curve C in space is said to be smooth in the xyz-plane if it has the parametric representation

$$\vec{r}(s) = x(s)\vec{i} + y(s)\vec{j} + z(s)\vec{k}, \qquad a \le s \le b,$$

where x, y, and z are continuous and have continuous derivatives for $a \le s \le b$. We assign a direction to C by choosing one of the two directions along C to be the positive direction (usually that of increasing s). Let $f(x, y, z)$, $g(x, y, z)$, and $h(x, y, z)$ be functions which are defined and continuous in a domain D of R^3. Then the line integral

$$\oint_C f \, dx + g \, dy + h \, dz \tag{B.1}$$

is said to be independent of path in D if, for every pair of endpoints A and B in D, the value of the line integral (B.1) is the same for all paths C from A to B in D. In other words, the value of the line integral depends in general on the endpoints A and B, but not on the choice of the path joining them.

The following theorem, which states that a double integral over a plane region can be replaced by a line integral over the boundary of the region, is known as *Green's theorem* and is fundamental in the theory of line integrals.

THEOREM B.1
Let R denote a closed bounded domain in the xy-plane whose boundary C consists of finitely many smooth curves. Let $P(x, y)$ and $Q(x, y)$ be functions which are continuous and have continuous first partial derivatives in R. Then

$$\oint_C P \, dx + Q \, dy = \iint_R \left(\frac{\partial Q}{\partial x} - \frac{\partial P}{\partial y} \right) dx \, dy. \tag{B.2}$$

If the line integral $\oint_C P\,dx + Q\,dy$ is independent of path in R, then

$$\oint_C P\,dx + Q\,dy = 0 = \iint_R \left(\frac{\partial Q}{\partial x} - \frac{\partial P}{\partial y} \right) dx\,dy, \tag{B.3}$$

which implies $\frac{\partial Q}{\partial x} = \frac{\partial P}{\partial y}$. Hence, there exists a function $\phi(x,y)$ defined in R such that

$$\frac{\partial \phi}{\partial y} = Q(x,y), \qquad \frac{\partial \phi}{\partial x} = P(x,y). \tag{B.4}$$

The converse of the above result does not hold without a further restriction, namely, that the domain R be *simply connected* [4]. In plain terms, a domain is simply connected if it has no "holes."

We now generalize Green's theorem where the double integrals are defined over a plane region to the case of surface integrals. This extension will require some basic facts about surfaces. A surface S in space is said to be a *smooth surface* if its normal vector depends continuously on the points of S. If S is not smooth but consists of finitely many smooth portions, then S is said to be a piecewise smooth surface. A domain D in R^3 is said to be simply connected if every simple closed curve in D forms the boundary of a smooth oriented surface in D.

One generalization of Green's theorem to R^3 takes the form of the following Stokes's theorem.

THEOREM B.2

Let S be a piecewise smooth oriented surface in space and let the boundary of S be a piecewise smooth simple closed curve C. Let $\vec{u}(x,y,z)$ be a continuous vector field, with continuous first partial derivatives in a domain that contains S. Then

$$\oint_C \vec{u} \cdot d\vec{r} = \iint_S (curl\ \vec{u} \cdot \hat{n})\,dA, \tag{B.5}$$

where $\hat{n}$ is the unit normal vector of S.

In addition, one can also generalize the above discussions on path independence for two to three dimensions. In fact, line integrals independent of path in R^3 are defined analogously as in R^2.

THEOREM B.3
Let $\vec{u}(x,y,z) = f(x,y,z)\vec{i} + g(x,y,z)\vec{j} + h(x,y,z)\vec{k}$ be a continuous vector field defined in a domain D of R^3. The line integral

$$\oint f\,dx + g\,dy + h\,dz$$

is independent of path in D if and only if there exists a scalar function $\Phi(x, y, z)$ *in D such that* $\vec{u} = \nabla\Phi$.

The above theorem immediately implies the following important result.

THEOREM B.4

Let $f(x, y, z)$, $g(x, y, z)$ *and* $h(x, y, z)$ *be continuously defined functions in a domain D of* R^3. *The line integral*

$$\oint f\,dx + g\,dy + h\,dz$$

is independent of path if and only if

$$\oint_C f\,dx + g\,dy + h\,dz = 0$$

for any simple, closed curve C in D.

From these theorems we may derive the following criterion.

THEOREM B.5

Let $\vec{u}(x, y, z) = f(x, y, z)\vec{i} + g(x, y, z)\vec{j} + h(x, y, z)\vec{k}$ *be a continuous vector field with continuous first partial derivatives in a domain D of* R^3. *If the line integral*

$$\oint f\,dx + g\,dy + h\,dz$$

is independent of path in D, then

$$\text{curl } \vec{u} = \vec{0}$$

everywhere in D. Conversely, if D is simply connected and $\text{curl } \vec{u} = \vec{0}$ *in D, then*

$$\vec{u} = \nabla\Phi$$

for some scalar function Φ.

We remark that a vector field $\vec{u}$ satisfying the condition $\text{curl } \vec{u} = \vec{0}$ in a domain D is termed vortex free or *irrotational* in D.

Finally, a second generalization of Green's theorem is the *divergence theorem* or *Gauss' theorem*, which also plays an important role in many theoretical and practical considerations.

THEOREM B.6

Let $\vec{u}(x, y, z)$ *be a vector field which is continuous and has continuous first partial derivatives in a domain D of* R^3. *Let S be a piecewise smooth oriented surface in D that forms the complete boundary of a closed bounded region R in D. Then*

$$\iint\limits_{S} \vec{u} \cdot \hat{n} \, dA = \iiint\limits_{R} \operatorname{div} \vec{u} \, dx \, dy \, dz,$$

where $\hat{n}$ is the outer unit normal vector of S.

References

[1] R. Buck, *Advanced Calculus*, McGrawHill, New York, 1978.

[2] R. Courant, *Differential and Integral Calculus*, translation by E.J. McShane, Interscience, New York, 1937.

[3] R. Courant and D. Hilbert, *Methods of Mathematical Physics, vol. II*, Wiley, New York, 1962.

[4] W. Kaplan, *Advanced Calculus*, Addison-Wesley Publishing Co., Inc., New York, 2003.

Index

For Product Safety Concerns and Information please contact our EU
representative GPSR@taylorandfrancis.com Taylor & Francis Verlag GmbH,
Kaufingerstraße 24, 80331 München, Germany

Printed and bound by CPI Group (UK) Ltd, Croydon, CR0 4YY
01/05/2025
01858518-0003